INTERNATIONAL ENERGY AGENCY

A GROUP STRATEGY FOR ENERGY RESEARCH DEVELOPMENT AND DEMONSTRATION

ORGANISATION FOR ECONOMIC CO-OPERATION AND DEVELOPMENT
PARIS 1980

Cover photo courtesy of the British Petroleum Co. Ltd.

Table of Contents

		Page
	Preface	7
I.	Introduction and Executive Summary	9
	A. Introduction	9
	B. The Project Approach	10
	C. Major Observations Resulting from the Analysis	12
	D. Recommended IEA RD & D Group Strategy	12
	E. Non-Technology Issues Affecting the Introduction of New Energy Technologies	15
	F. Use to be Made of the IEA Group Strategy	15
II.	Systems Analysis: The Analytical Tool for Strategy Development	17
III.	Summary of Systems Analysis Results	27
	A. Introduction	27
	B. Energy System Overview	27
	1. Oil and Gas	35
	2. Conservation	35
	3. Coal	36
	4. Nuclear Power	37
	5. Renewable Energy Sources	38
	6. Assessment of Overall Energy System Results	39
	C. The Energy System Cost/Security/Technology Acceleration Trade-Offs	40
	D. Major Observations Resulting from the Analysis and Their Implications for an IEA RD & D Group Strategy	41
	E. Effects of Uncertainties	46
IV.	Recommended Group Strategy	49
	A. Introduction	49
	B. Criteria for Establishing Priorities	49
	C. Applying the Criteria	51
	D. Recommended Group Strategy	56
V.	Environmental Considerations	59
VI.	Some Non-Technology Policy Issues Affecting The Introduction of New Energy Technologies	61
VII.	Use of the Recommended Group Strategy and Future Activities	63
	A. Use of the Group Strategy	63
	B. Future Activities	63
	Project Bibliography	65
	Annex I : IEA Principles for Energy Policy Adopted by the Governing Board and Excerpt from IEA Ministerial Level Governing Board Meeting Communique of 22nd May, 1980	67
	Annex II : Technology Characterization Summary	69
	Annex III: Overview of Systems Analysis Results	77

List of Figures

		Page
1.	Group Strategy Generic Technology Priorities and Indicative Actions	14
2.	Energy System in Markal Model	18
3.	Generic Technologies and Examples of Processes	19
4.	Oil Price Assumptions	22
5.	Primary Fuel Price Assumptions - Germany	23
6.	List of Scenarios	25
7.	Average Annual Undiscounted Capital Investment Index Under Four Scenarios	31
8.	Sources and Uses of Certain Energy Forms in the Accelerated Security Scenario (SP-4/PREM-2)	32
	a. Sources of Gas	32
	b. Sources of Electricity	32
	c. Sources of Liquid Fuels	33
	d. Uses of Liquid Fuels	34
	e. Sources of Heat in the Residential/ Commercial Sectors	34
9.	Energy System Cost/Security/Technology Acceleration Trade-Off Curves for 15 IEA Countries	40
10.	Energy System Cost/Security Trade-Off Curves Illustrating Differences Among Three Countries	42
11.	Net Conventional Oil Imports Under Four Scenarios	43
12.	Group Strategy Priority Criteria	50
13.	Technology Priorities in a Minimum System Cost Strategy	52
14.	Technology Priorities in a High Security of Energy Supply Strategy	54
15.	Group Strategy Generic Technology Priorities	55
16.	Group Strategy Indicative Actions for Generic Technologies	57
17.	Group Strategy Principal Environmental Factors	60

List of Tables

		Page
1.	GDP Growth Rate Assumptions	21
2.	Average Annual Growth Rate of Total Primary Energy (TPE) Indicated by the Minimum System Cost Scenario (PS-1)	22
3.	Energy Data Indicated for 15 Countries: Minimum System Cost Scenario: Non-Accelerated (PS-1)	27
4.	Energy Data Indicated for 15 Countries: Minimum System Cost Scenario: Accelerated (PS-4)	27
5.	Energy Data Indicated for 15 Countries: Non-Accelerated Security Scenario (SP-1/PREM-2)	28
6.	Energy Data Indicated for 15 Countries: Accelerated Security Scenario (SP-4/PREM-2)	28
7.	Conventional Sources of Primary Energy in Two Scenarios	28
8.	New Sources of Primary Energy in Two Scenarios: Percentage Contribution to Total Primary Energy Supply	28
9.	Relative Use and Average Annual Growth Rate of Primary Energy: Minimum System Cost Scenario (PS-1)	30
10.	Relative Use and Average Annual Growth Rate of Primary Energy: Accelerated Security Scenario (SP-4/PREM-2)	30
11.	Impacts of New Conservation Technologies in the Minimum System Cost Scenario	35
12.	Coal Utilization in Two Scenarios	36
13.	Total Net Coal Import Requirements for 15 IEA Countries	37
14.	Uranium Requirements in Two Scenarios and WOCA Resources	38
15.	Oil Imports as a Percentage of Total Oil Requirements in Two Scenarios	47
16.	Energy Contributions by Technology in Four Minimum System Cost Scenarios	51
17.	Energy Contributions by Technology in Four Accelerated Security Scenarios and a Maximum Renewables Scenario	53

PREFACE

Early in 1976 the IEA Committee on Energy Research and Development (CRD) initiated a programme of energy technology systems analysis and strategy development. This document is a report of that effort. It presents a strategic view of new energy technologies from the standpoint of the overall energy needs and prospects for the IEA countries as a group. It thus does not necessarily portray the importance of any individual technology for any one country. Similarly, it does not portray a recommended strategy for any individual country.

The CRD wishes to express its appreciation for the great contributions to the programme by the Kernforschungsanlage Jülich GmbH (Germany) and Brookhaven National Laboratory (U.S.), by the many individuals from participating countries and the Commission of the European Communities who worked on the multinational teams at those laboratories, and by the Steering Group for Energy R & D Systems Analysis, which guided the systems analysis portion of the programme. This latter effort served as a vital input to the strategy development. While CRD provided policy guidance to the systems analysis, the details of that work and its results are the responsibility of the Steering Group and the Laboratory teams.

Chapter I

INTRODUCTION AND EXECUTIVE SUMMARY

A. INTRODUCTION

The oil embargo of 1973/74 signalled a fundamental change in the ability of the industrialized nations to chart their own economic destinies and to guarantee the economic security of their citizens. Only major wars and economic depressions have directly affected so many people in the world's oil importing nations. The fragile nature of the world's energy supplies was demonstrated once again with the substantial disruptions and continuing uncertainties in the oil market starting with the political upheaval in Iran early in 1979.

As IEA nations have taken measures to accommodate to the new energy circumstances, they have recognized the need for action to reduce their vulnerability in the near term, and that additional actions are necessary to pave the way for more lasting, long-term solutions. The needed actions involve strong efforts towards more rational and efficient use of energy (conservation) and the use of supply alternatives to rapidly depleting and increasingly costly oil. While many such actions can make use of existing technologies, it has been recognized as a part of overall energy policy that in the long term the development of new and improved technology is required, both for conservation and energy supply.

All IEA countries, therefore, are engaged in energy research, development and demonstration (RD & D) programmes. In 1980 the twenty-one IEA governments are expected to spend over $7 billion for RD & D on new energy technologies. Industry is also devoting large amounts of funds to this field. In each country, those programmes were undertaken on the basis of a national assessment of needs for technological advancement and were strongly influenced by national circumstances. However, such an approach to programme planning may not adequately consider the total, collective need for, and potential contribution of, new energy technologies. It may, therefore, fail to address that need in terms of overall programme size, scope and balance. Accordingly, early in 1976 a number of IEA countries agreed to collaborate within the IEA Committee on Energy Research and Development (CRD) in the development of an overall IEA research, development, and demonstration strategy, and to join in a co-operative analytical effort as a basis for the strategy.

An IEA energy RD & D strategy is fundamentally different from individual national RD & D strategies. It is a view of the energy technology needs of the IEA nations as a group, and an attempt to answer the questions:

- — Which new energy production, conversion and end-use technologies will IEA countries as a whole need over the next several decades, and what energy contribution can be expected of them?
- — How might the totality of IEA energy RD & D programmes be structured in order to maximise the chances of having the technologies when they are needed?

Since the IEA countries are not a single nation and there are many individual country considerations that influence national RD & D programmes, the IEA group strategy is not intended to be, nor can it be, identical to individual national strategies and programmes. Rather, it is intended to be a guide for individual national energy RD & D programmes taking into account the collective needs of IEA countries.

More specifically, the purposes of the IEA RD & D group strategy are:

1) To provide an assessment of the likely relative importance of individual technologies for the IEA nations as a group;

2) To obtain estimates and from these develop targets for the energy impact to be achieved by

new and improved technologies during the latter part of the century;

3) To provide a tool for:
   — developing and assessing national RD & D policies and plans;
   — achieving the most effective relationship between the RD & D activities of the Member countries;

4) To identify non-technology policy issues which can affect the ability of new technologies to contribute to energy requirements.

The first, preliminary phase of this project was completed in the spring of 1977 and a report published in September 1977, entitled *Towards an International Strategy for Energy Research and Development*. That report, as well as studies by others since then, stressed the high probability of a fundamental imbalance between energy supply and demand during the remainder of this century.

Since that preliminary report, a thorough and in-depth review of the prospects for new energy technologies through the year 2020 was undertaken. Fifteen countries participated in the project. In 1977 they accounted for 95 per cent of the total primary energy consumption of the 21 IEA countries, and 95 per cent of the total IEA oil consumption (94 per cent of IEA oil imports).*

The analytical effort was performed by multinational teams assembled at the Kernforschungsanlage Jülich GmbH (KFA), Jülich, Germany, and the Brookhaven National Laboratory (BNL), Upton, N.Y., USA, under the direction of a Systems Analysis Steering Group established by the IEA Committee on Energy Research and Development. Participating in this effort were: Austria, Belgium, Canada, Denmark, Germany F.R., Ireland, Italy, Japan, New Zealand, Norway, Spain, Sweden, Switzerland, the United Kingdom, the United States, and the Commission of the European Communities. The Netherlands participated in the initial phase of the work.

Using the quantitative results obtained from the systems analysis work as a starting point, an IEA group RD & D strategy was developed. For the purposes of this strategy the basic energy policies which have been followed are the Principles for Energy Policy adopted by the IEA Governing Board at Ministerial level in October, 1977. They are set forth in Annex I.

The relationship between an RD & D strategy and a total energy strategy is an iterative one. RD & D begins as an input to overall energy policy, and once policy decisions have been made regarding its appropriate role, then a more definite RD & D strategy can be developed. This relationship is present on a national level as well as internationally and, among other factors, must be recognized when this IEA group strategy is used as a guide for national RD & D efforts.

While the focus of the effort was on the potential use of new energy technologies in IEA countries, these technologies can be of broader benefit. Directly and indirectly they can be of importance to developing countries whose need for energy is expected to increase at a more rapid rate than the industrial countries and most of whom are severely burdened as the cost of imported oil increases. Some of the technologies can also benefit oil producing countries by increasing the yield from their resources.

The results of the project as presented in this report are only a beginning. This IEA strategy cannot answer all relevant questions about energy RD & D for all time. For example, it only treats the potential of families of technologies relative to one another; it does not address strategies for the development and use of individual technologies. It does not address in detail the problems - industrial, institutional and social - which can be encountered when introducing a new technological system into commercial use; this subject is of considerable importance and is on the continuing agenda of the CRD.

B. THE PROJECT APPROACH

The Process

Two distinct steps were involved in the development of the strategy: (1) a detailed and systematic analysis of the possible energy contributions of selected new technologies over a forty-year period under a variety of assumptions; and (2) the application of strategic planning judgements to the analytical results. While the analysis included both conventional and new technologies, the strategy project is focussed on new technologies.

In the analytical phase, because of the very large number of variables involved in estimating the degree to which new technologies would compete

* In 1977, IEA countries used 76 per cent of the energy and 74 per cent of the oil consumed world wide outside the centrally planned economies.

in the energy market over a long time horizon widely-used energy modelling techniques were employed. The model simulated the entire energy system of a country over the forty-year time period from primary energy supply, through all conversion processes, to end-use (Figure 5). Each country's system was described, and the economic and technical characteristics of 27 generic production, conversion and end-use technologies (Fig. 6) were estimated and used in a manner which preserved commonality of these estimates from country to country, while at the same time taking account of differences in national circumstances (Annex II).

In both the analytical work and the strategy development two major policy thrusts were examined - the need to minimize the overall cost of the energy system in order to minimize the effect of rising energy costs on societal goals and the need to minimize the imports of oil to enhance security of energy supply. Trade-off studies of these two thrusts were also made to determine their interactions.

Assumptions

In an analytical effort of this nature, assumptions entering the analysis significantly shape the results. The most critical assumptions are the world prices of oil, the prices and costs of domestic fuels, the availability and producibility of fuel resources, economic and associated energy growth rates, and the relative costs of new technologies.

With respect to future energy supply and demand, the assumptions used were based upon other studies. The key assumptions were:

- — Modest and gradually declining economic growth rates (Table 6);
- — Continually increasing oil and other real energy prices (Figures 7 and 8);
- — Substantially reduced growth of end-use energy requirements through increased energy use efficiency.

Recognizing the major uncertainties inherent in projecting future events, studies were made of the effects of varying some of the assumptions - future oil prices, certain possible societal preferences (i.e., restrictions on use of fossil or nuclear fuels, and preferential use of renewable resources), and of the possible effects on oil imports and system costs of different rates of technology development and introduction (current rates versus making special efforts to accelerate the introduction of new technologies). Other uncertainties are also present, such as the size and availability of natural resources, the course of future economic development, and the results of RD & D efforts themselves.

Limitations of the Approach Employed

The many unknowns and uncertainties inherent in a study of this nature dictate that caution be applied in viewing the numerical results of the systems analysis work as presented in this report. The essence of energy RD & D strategy development has been to gain the insights that they offer without attributing to the numbers a precision that they do not have.

The analytical effort, while extensive, nevertheless did not explicitly address a number of issues which have relevance in estimating the future potential of technologies.* In making judgements to arrive at its recommendations, CRD attempted to take these limitations into account, but it is useful to note the most important of them:

- — The projected costs of the individual technologies, particularly as they compare with other technologies to supply an energy need, strongly influence the results. While an attempt was made to use the best information available, there are inherent uncertainties in such projections. Comprehensive sensitivity studies of the effect of changes in estimates of the cost and other characteristics of individual technologies were not performed.
- — The maximum rate at which a new technology can be implemented once it becomes commercial has been estimated. The best available information was used in these estimates, although in many cases they are not the result of detailed studies.
- — While questions relating to social and environmental impacts were identified, it has not been possible to quantify those impacts.
- — It was assumed that trade in energy fuels and technologies among IEA countries was unrestricted. To the extent this assumption proves to be incorrect, the strategy of individual countries and of the IEA group could be affected.

* It should be noted that a large body of analytical results were developed in the course of this project which are not directly reflected in the strategy development. This information remains available for use in strategy refinements.

— The analysis is based upon one set of demand projections. Within broad limits, it is not expected that variations in demand would affect the relative importance of the technologies studied, with the important exception of major changes in energy use patterns.

— Energy conservation technologies were only partly treated explicitly; much of their effect was included in demand projections. Therefore, the potential contributions of conservation-related technologies are understated in the results.

C. MAJOR OBSERVATIONS RESULTING FROM THE ANALYSIS

The systems analysis effort led to the following major observations of importance in RD & D planning:

1. During the 1980-2020 period the requirements for imported conventional oil are projected to decline. During the first twenty years the decline is mainly caused by conservation and expansion of existing technologies. During the second twenty years new liquid fuels technologies begin to make a significant contribution.
2. At today's pace of energy RD & D and commercialization efforts, the resulting rate at which imported oil declines would continue to leave IEA too vulnerable to supply disruptions. The contribution of new energy technologies before the year 2000, therefore, may not adequately reduce demand for oil unless their introduction is accelerated.
3. The impact of new conservation technologies can be very significant in meeting IEA objectives. The utilization of energy conservation technologies, despite the only partial treatment made in this study, approximates that of new supply technologies.
4. The major growth in primary energy occurs in coal and nuclear power in virtually all countries. High priority must, therefore, be attached to those RD & D efforts which support and facilitate the use of these energy sources.
5. Electricity growth continues to exceed total energy growth rates in all cases studied, although the rate of growth slowed down through each successive ten-year interval.
6. Some new renewable technologies begin to make significant energy contributions before 2000; however, non-renewable technologies are called upon to provide the major fraction of energy needs even in the case of accelerated efforts to introduce renewables.

D. RECOMMENDED IEA RD & D GROUP STRATEGY

In view of the possibility of substantially reducing oil imports through accelerated development and deployment of promising technologies at relatively modest additional energy system costs, *it is the view of CRD that a strategy of high security of energy supply be adopted.*

The Group Strategy has the following elements:

— The pace of technology considered essential if IEA energy policy objectives are to be realized;

— The priorities which individual generic technologies should have;

— Indicated actions proposing the next necessary and possible steps with respect to RD & D and commercialization of generic technologies; and

— Environmental considerations.

The CRD recommendations on these and related issues are summarized below.

1. The pace of new technology development and introduction should be accelerated

The efforts of IEA countries to develop and bring new energy technologies to general commercial use must receive significantly higher priority than they do at the present time. This involves not only public and private sector expenditures for energy RD & D, but also strenuous efforts to overcome barriers to the commercialization of new energy technologies.

A rough indication of the differing effects on oil imports between acceleration and non-acceleration of new energy technologies may be seen in the figure below.* It may not be possible to achieve the precise situation shown by the Accelerated Case curves, but the analysis indicates that acceleration of new energy technologies can have major oil

*Figure 13 shows sources and uses of certain energy forms under the accelerated security scenario (SP-4/PREM-2).

IEA ENERGY RD & D STRATEGY PROJECT

NET CONVENTIONAL OIL IMPORTS UNDER FOUR SCENARIOS

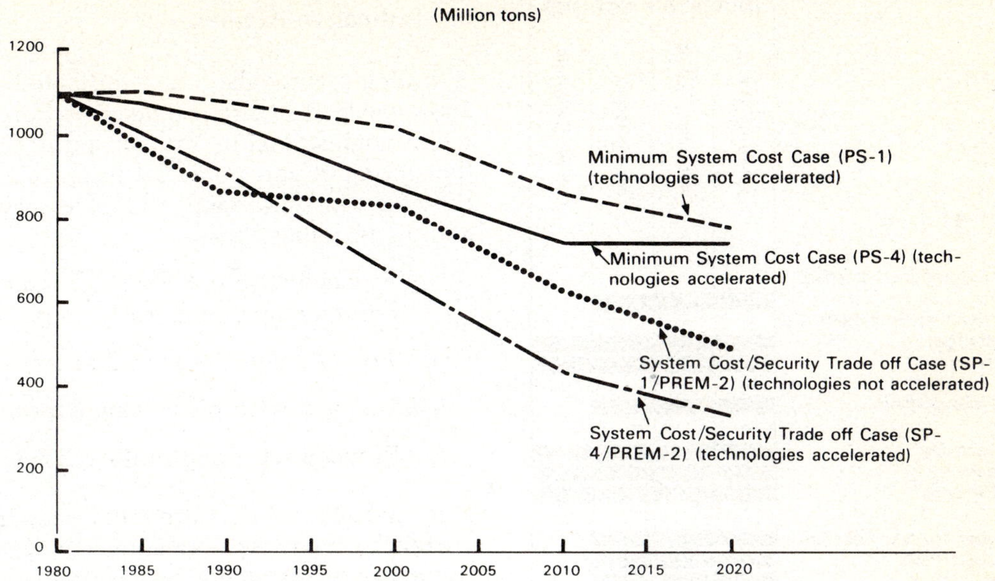

saving effects. Although it has not been possible to estimate the cost of achieving acceleration, the analysis to date indicates that acceleration does significantly reduce total energy system costs. This results from the earlier and increased application of supply and conservation technologies that substitute for more expensive oil that would otherwise be required. It will be necessary for each country to assess the costs and benefits of acceleration on a case-by-case basis.

2. Generic Technology Priorities

To convert the results of the analyses performed in the project into a meaningful strategy, it is necessary to take the insights gained from the analysis and apply additional judgements which must guide strategic planning. Three principal criteria were used to place the 27 generic technologies into priority categories:

i) the potential amount and timing of each technology's contribution to energy supply or energy conservation;

ii) the degree to which the current state of knowledge makes it possible to estimate that the technology can successfully achieve the economic and technical goals assumed for it; and

iii) the importance that the technology may have as a "hedge" against uncertainties.

On the basis of these criteria, CRD placed the technologies into four priority categories as shown in Figure 1, with no rank order intended within each category. (There are, clearly, a number of processes or products within each generic category that have varying promise; it is not intended that all are of equal priority.)

In applying the above priorities, it is important to recognize that:

— In view of the uncertainties inherent in all such projections of the future, even the lower priority technologies may prove to have significant value and therefore require proper attention;

— Priority ranking should not be interpreted as translating directly into relative amounts of RD & D money to be spent among the individual technologies. Rather, it is intended to provide guidance in determining which technologies should have priority call for funding, considering their state of development. Thus, a high priority technology which is at an early exploratory phase may appproprately have a lower budget than a lower priority technology which is at the stage of demonstration. However, if the higher priority technology has need for additional funds to carry out a significant next step, it should have prior call on such funding; and

Figure 1

IEA ENERGY RD & D GROUP STRATEGY

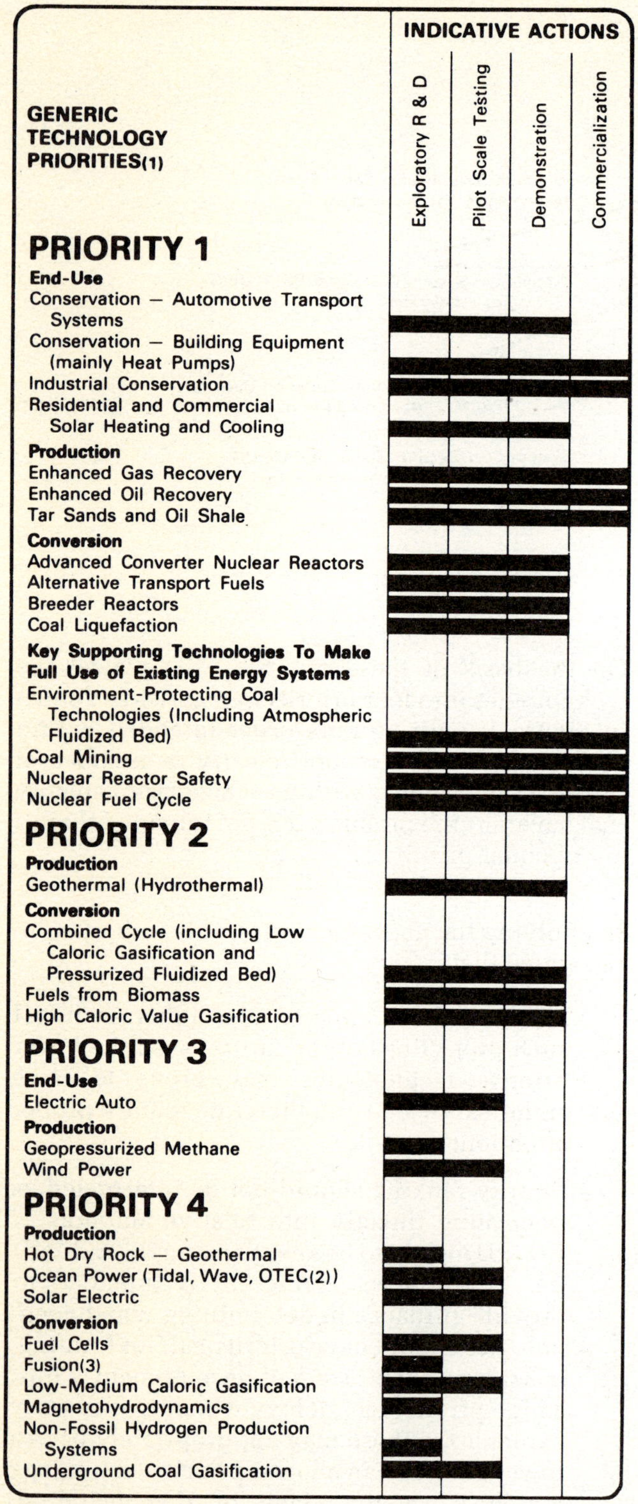

(1) The priorities apply to IEA countries as a group and not to any individual country.
(2) Ocean Thermal Energy Conversion.
(3) Fusion is ranked in Priority 4 due to its late availability (2020). However, taking into account its potentially large contribution and long lead time, sufficient efforts should be made to develop this technology.

— The priorities apply to IEA countries as a group and not to any individual country.

3. Indicative Actions

A strategic view also must indicate the near-term actions which should be taken with respect to the technologies, taking into account their state of development and potential importance. Figure 1, therefore, also indicates the necessary or possible actions in four stages:

i) Proceed with the steps necessary to bring to commercial use as early as practicable;

ii) Proceed with large scale demonstrations;

iii) Proceed with pilot scale experiments;

iv) Proceed with exploratory R & D.

It is emphasized that the table only indicates the current or next step warranted. In each case, work in the preceding stages may well need to continue.

Further, since the technologies shown are generic, usually encompassing a number of processes, the stage of development of the most advanced process(es) is portrayed. It is not intended that each individual country will necessarily find the development at the stage indicated or will find it appropriate to engage in any or all the stages; some countries may not be working in the area at all, or may contribute only a portion of the total effort needed to explore and develop the particular technology.

4. Environmental Considerations

Virtually all energy technologies have some environmental impact (Figure 17). Given the importance of adequate protection of the environment, it is an essential part of the group strategy that the environmental impacts be understood and taken into careful account at all stages of technology development, and prior to deployment. While an attempt was made to include those impacts in the analysis, it became clear that it was not possible to identify an acceptable means of relating quantitatively the various environmental effects of the 27 technology categories. However, there has been a limited, explicit inclusion of this factor in the above technology priority categories in two ways: (1) environmental control costs are included in some of the individual technology areas; and (2) certain known research and development issues are taken into account in the above statement of priorities.

5. Periodic Re-examination Required

It must be recognized that an energy RD & D strategy is evolutionary. The strategic view presented here is based upon information and insights available today. Information and insights change with time so that the strategy must be periodically re-examined in due course. The CRD intends to do so.

E. NON-TECHNOLOGY ISSUES AFFECTING THE INTRODUCTION OF NEW ENERGY TECHNOLOGIES

The deployment of new technologies can be affected by a wide variety of non-technological issues. There must be adequate attention to such issues in the course of technological development if the potential of new technologies is to be realised. Among the most important are:

— **Rational Pricing Signals.** Artificially reducing the prices of different forms of conventional energy, for example, as a result of selective price controls, can impede the introduction of new technologies as well as conservation.

— **Foreign Trade Policies.** Trade in primary energy as well as in energy technologies and know-how needs to be facilitated.

— **Social Impacts.** Governments will need to anticipate, consider and minimize potentially troubling social impacts that may result from the large-scale introduction of some new energy technologies.

— **Supporting Resource Availability.** Action may be needed regarding resources such as skilled labour, raw materials, infrastructure, land and water availability. This is particularly important in the case of expanded coal use.

— **Regulatory and Industrial Structure.** Often government regulations and industrial structure, which were designed around conventional systems, fail to accommodate the character of new technologies.

F. USE TO BE MADE OF THE IEA GROUP STRATEGY

The CRD, as part of its process of reviewing national energy RD & D programmes, encourages Member countries to develop long range national energy RD & D strategies. The group strategy is intended to be considered by individual governments to provide insights for use in the development of national strategies, and by the CRD as it considers the totality of RD & D efforts of IEA countries.

1. At the National Level

a. Member countries are urged to accelerate their efforts to develop and bring to commercial use important new energy technologies.

b. As national energy RD & D strategies are developed and evolve, the Group Strategy technology priorities should be one of the criteria used in setting national RD & D priorities. Other criteria will, of course, include the individual country's natural and economic resources, technological and industrial structure and energy policy objectives. In its periodic reviews of national energy RD & D programmes, the CRD will consider the insights provided by this group strategy in evaluating the strategies of individual countries, taking into account relevant national circumstances such as the following:

— national energy policies

— existing technological capability

— natural resources

— industrial development objectives

— overall economic resources.

2. At the International Level

a. Using information collected in the course of its annual review of Member countries' energy RD & D programmes, the CRD intends to compare the aggregate of national energy RD & D efforts with the Group Strategy and to make recommendations accordingly.

b. In its collaborative project activities, the CRD will consider the importance of the technologies as portrayed in the strategy, recognizing that there are additional factors which influence the potential for, and value of, such collaboration.

Chapter II

SYSTEMS ANALYSIS: THE ANALYTICAL TOOL FOR STRATEGY DEVELOPMENT

A. GENERAL

The development of the IEA energy RD & D strategy involved two basic steps:

i) **The analytical phase:** Comprising an assessment of the likely energy contributions to be made by new technologies over the period of interest; and

ii) **The strategy development phase:** Covering an assessment of the implications of those contributions, coupled with other relevant considerations, to define RD & D actions to be taken now and in the future.

For the first of these steps, the analytical phase, a set of studies was made of the extent to which new energy technologies might be used over the next forty years under differing assumptions of the future (scenarios). The studies were performed by the systems analysis teams at KFA and BNL.

Each individual scenario study was based upon one set of assumptions and therefore depicted only one possible future outcome. It is not necessarily of high probability. It is certainly not a forecast. Taken together, the several studies are designed to provide insights on the performance of new energy technologies within the total energy system. In addition, the analysis provides observations on the patterns, costs and quantities of fuel substitutions, oil imports, electric energy demand and supply, and the possible growth of coal, gas, nuclear and renewable energy, and conservation. Such studies were made for each of the fifteen participating countries.

For these studies, it was necessary to:

i) Develop an analytical model of the way in which energy is used;

ii) Forecast, as well as the present state of knowledge permits, the characteristics (technical, economic, and environmental) and estimate the maximum practical rate of deployment of the technologies which might be developed through RD & D efforts;

iii) Make assumptions (or ranges of assumptions) as to future end-use energy requirements, availability of fuels, and energy prices;

iv) Select a set of possible policy options of importance to decision makers;

v) Determine, through analysis, the possible energy contributions of each technology under the alternative sets of assumptions and policy options.

The analyses illustrate the *relative* contributions of new technologies and energy use patterns towards the achievement of varying objectives. Thus, they are an aid to thinking about and understanding options for the future, and assisting in evaluating the consequences of alternative policy choices. It is the understanding of the behaviour of the energy system under different assumptions and its sensitivity to changes in assumptions, not the final computer tabulation of numbers, that provides the basis for RD & D strategies.

The following sections of this Chapter describe in greater detail the analytical approach employed in this project. The use of the analyses in strategy development is set forth in later chapters.

B. THE ANALYTICAL APPROACH

1. The Development of the Analytical Model

The Market Allocation (MARKAL) Model was specifically developed for this project, building upon the best features of existing models at BNL and KFA. The model is a multi-time-period linear programming model, and was applied to the period 1980-2020, in five-year intervals. It covers the entire flow of energy from primary sources through the transformation sector to final energy and through the utilization sector to useful energy (see Figure 2). In the working of the model, these energy demands can be met by the optimum combination of existing energy forms or through the introduction of new technologies. The model may be used to portray the optimum way in which different forms of energy can be used to meet the many different energy needs against a particular objective, e.g. minimum total discounted energy system cost, minimum cumulative oil imports, or the maximum use of renewable resources.

Figure 2

IEA ENERGY RD & D STRATEGY PROJECT

ENERGY SYSTEM IN MARKAL MODEL

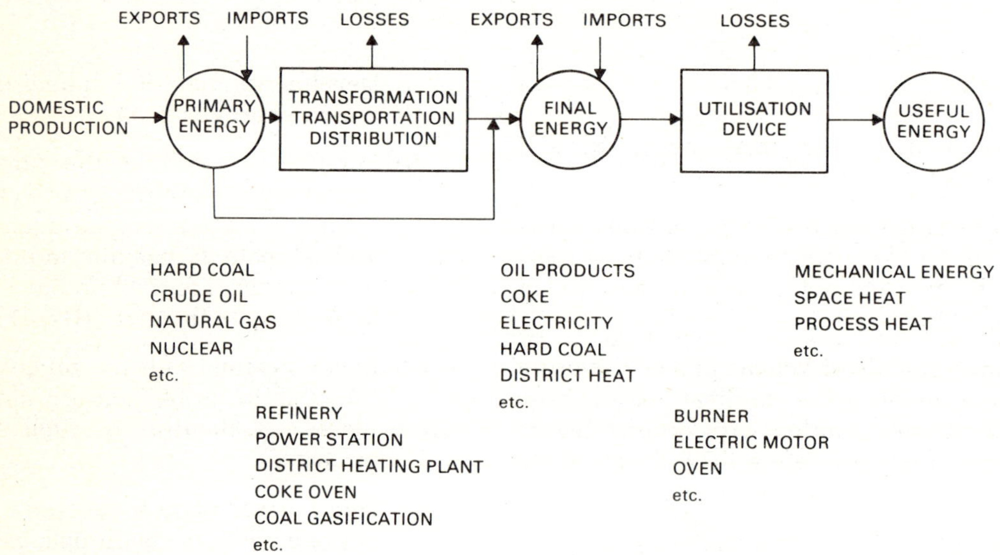

MARKAL can incorporate a large number of existing or new energy technologies representing either energy production, conversion or end-use systems. An important feature in MARKAL is the ability to include conservation and end-use technologies in the same manner and level of detail as production and conversion technologies. The model selects the mix of production, conversion and end-use technologies (subject to the date and rate of availability and the constraints specified in the characterization) that best satisfies a set of energy demands under the conditions defined in each scenario.

The documentation of the MARKAL Model, explaining the methodology and the details of the application, are available in separate BNL and KFA publications.

2. Input Data for the MARKAL Model

The key input data for MARKAL are:

— a quantitative description of each existing and new technology and technical limitations in the growth rate of deployment of these technologies;

— energy demand for each demand subsector for each time period studied;

— the costs of natural fuel resources and technical limitations on the growth rate of usage of such resources.

The manner in which these input data were developed is described in the following sections.

a) The Characterization of the Technologies. The strategy project included 27 families of new technologies in the analysis (Figure 3). They were intended to cover the major new technologies which are believed to have the potential of making a contribution to meeting future energy needs, though not every new technology of interest was covered.

Figure 3

IEA ENERGY RD & D STRATEGY PROJECT

GENERIC TECHNOLOGIES AND EXAMPLES OF PROCESSES EXAMINED IN THE SYSTEMS ANALYSIS

End-Use

CONSERVATION-AUTOMOTIVE TRANSPORT SYSTEMS
- Private Auto, Improved Frame and System
- Private Auto, Improved Diesel
- Private Auto, Stirling Engine

ALTERNATIVE TRANSPORT FUELS
- Private Auto, Methanol Fuels
- Compressed Natural Gas

CONSERVATION BUILDING (EQUIPMENT)
- Electric Heat Pump
- Gas Heat Pump
- Seasonal Low-Temperature Heat Storage, Underground

ELECTRIC ATUO
- Battery-Powered Auto

INDUSTRIAL CONSERVATION
- Solar to Process Heat
- Atmospheric Fluidised Bed Industrial Boiler for Process Heat

RESIDENTIAL AND COMMERCIAL SOLAR HEATING
- Space Heat
- Water Heat
- Air Conditioning

Production

ENHANCED GAS RECOVERY

ENHANCED OIL RECOVERY

GEOPRESSURIZED METHANE

GEOTHERMAL/HYDROTHERMAL/HOT DRY ROCK
- Dry Geothermal (Hot Rock) Power Plant
- District Heating Plant, Geothermal

OCEAN POWER
- Wave Central Electric Power Plant
- Ocean Thermal Gradient Electric Power Plant

SHALE AND TAR SANDS

SOLAR ELECTRIC
- Decentralized Solar Photovoltaic
- Central Solar Thermal Electric
- Central Solar Photovoltaic Plant

Figure 3 *(cont'd)*

IEA ENERGY RD & D STRATEGY PROJECT

GENERIC TECHNOLOGIES AND EXAMPLES OF PROCESSES EXAMINED IN THE SYSTEMS ANALYSIS

WIND POWER
- Wind Turbines, Central Electric Power Complex
- Local, Wind Electric Generator

Conversion

ADVANCED CONVERTER REACTORS
- ATR Nuclear Power Plant
- CANDU Nuclear Power Plant, Enriched Fuel
- HTR Nuclear Power Plant
- Nuclear Process Heat, VHTR

BREEDER REACTORS

COAL LIQUEFACTION
- Hydrogenation
- SRC-2 Process
- Methanol

COMBINED CYCLE
- Hard Coal Combined Cycle Power Plant
- Pressurized Fluidized Bed Power Plant
- Atmospheric Fluidized Bed Power Plant

FUEL CELLS
- Gas Fuel Cell

FUELS FROM BIOMASS
- Biomass Steam Electric Power Plant
- Methanol from Wood
- Gas from Wastes

FUSION
- Nuclear Fusion Power Plant

HIGH CALORIC GASIFICATION
- High Btu Gasification, LURGI-Slagging
- Nuclear Hydrogasification

LOW-MEDIUM CALORIC GASIFICATION
- Hard Coal, Medium Btu Gasification
- Hydrogen Production from Hard Coal

MAGNETOHYDRODYNAMICS
- Coal MHD Electric Power Plant

NON-FOSSIL HYDROGEN SYSTEMS
- Hydrogen Production by Electrolysis of H2O

UNDERGROUND GASIFICTION
- Underground Gasification with Combined Cycle Electric Power Plant

Within each technology family, each country was able to include one or more specific processes. Each of the specific technological processes was described by a set of quantitative data required for MARKAL input. The data included capital investment, operating and maintenance costs, efficiencies, type of fuel inputs and outputs, and technical lifetimes. For new technologies, the date of commercial availability was specified and the maximum feasible technical rate of introduction estimated. For the existing stock of technologies, residual lifetimes were given. Examples of some characterization data, including availability dates for many of the new technologies, are shown in Annex II.

The required data were assembled over a two and one-half year period using a variety of expert sources in industry, government agencies, research laboratories and published reports. Often a range of estimates exist, even among experts. These ranges mainly reflect the inherent uncertainties in the data. The systems analysis teams attempted to reconcile different opinions by investigating the underlying causes. Often the differences could be explained, e.g. variations in design, cost differences due to unit size, variations in pollution abatement equipment, etc. But in some cases, best judgement choices had to be made among conflicting data.

To promote consistency of results, reference characteristics were established for common use by all countries. Variation from the common characteristics was permitted when it was necessary to account for differences unique to a particular country. (Because of the national differences that exist in the form and selection of end-use technologies, uniform characterizations of such technologies were less possible.)

A comprehensive documentation of the characteristics of conversion and end-use technologies, which served as a basis for the reference characterizations, are contained in the Energy Technology Data Handbook. Reference values, as well as variations from these used by individual countries, for the production and conversion technologies, are documented in detail in the Technology Review Report. (see Project Bibliography)

Table 1

IEA ENERGY RD & D STRATEGY PROJECT GDP GROWTH RATE ASSUMPTIONS USED IN THE ANALYSIS[1] AND ACTUAL 1966/7 TO 1976/7 ANNUAL AVERAGES[2]

	Average 1966/7 to 1976/7	1980-1985	1985-1990	1990-1995	1995-2000	2000-2005	2005-2010	2010-2015	2015-2020
AUSTRIA	4.6	2.4 →		1.9 →		1.5 →			1.3
BELGIUM	4.5	3.0 →				2.5 →		2.0 →	
CANADA	4.7	3.4 →				1.6 →			
DENMARK	3.6	3.5	3.0 →			2.1 →			
GERMANY	3.6	3.5	2.5 →			2.0 →		1.5 →	
IRELAND	4.3	5.0 →		4.0 →					
ITALY	3.9	3.2	2.5 →			1.9 →			
JAPAN	7.8	6.0	4.0 →			3.0 →			
NEW ZEALAND	2.9	3.1	3.0	2.8	2.7	2.6	2.5	2.3	2.25
NORWAY	4.6	3.8 →				2.5 →			
SPAIN	5.2	5.0	4.5 →			3.5 →			
SWEDEN	2.4	2.8 →				2.2 →			
SWITZERLAND	2.1	2.5 →				1.0 →			
U.K.	2.1	3.0	2.5 →			1.9 →		1.6 →	
U.S.	2.8	4.0	3.1 →			2.4 →			

1. Assumptions were prepared by project staff, and do not represent government projections.
2. Source : OECD *Economic Outlook,* July 1980.

b) The Setting of Key Economic and Energy Demand Assumptions.

It was not the purpose of this project to provide yet another study of future energy demand; such projections are used in this case only as a framework for examining the possible contribution of new technologies. Accordingly, economic growth, energy demand and fuel price projections were established by drawing upon other studies and expert judgement.

Each country established an economic growth rate within the guidance of growth ranges provided by the strategy project (Table 1). The values for the individual countries have been checked against other OECD/IEA projections and have been found to be generally consistent with these figures, which, however, do not go beyond 2000. The growth rates in primary energy emerge as output from the analysis and are shown in Table 2.

Table 2

IEA ENERGY RD & D STRATEGY PROJECT AVERAGE ANNUAL GROWTH RATE OF TOTAL PRIMARY ENERGY (TPE) INDICATED BY THE MINIMUM SYSTEM COST SCENARIO (PS-1)*

Percent

Country	1980-1990	1990-2000	2000-2010	2010 2020
AUSTRIA	1.6	1.4	1.2	1.2
BELGIUM	1.9	1.5	1.0	0.7
CANADA	3.0	2.4	2.5	2.7
DENMARK	1.8	1.8	1.3	1.3
GERMANY	1.6	1.1	1.0	0.8
IRELAND	5.5	3.4	4.0	2.3
ITALY	2.0	1.6	0.8	0.9
JAPAN	3.9	2.4	2.1	1.6
NEW ZEALAND	3.0	2.8	2.2	2.2
NORWAY	3.0	2.8	2.2	2.2
SPAIN	2.7	2.1	1.5	1.1
SWEDEN	0.6	0.9	0.9	1.4
SWITZERLAND	2.0	2.3	0.8	0.9
UNITED KINGDOM	1.2	1.1	1.3	1.5
UNITED STATES	1.6	1.9	1.9	1.6

* TPE is calculated in the MARKAL model.

Each country established its energy demands to be satisfied over time. Where possible the demands were expressed in terms of demands for services, such as space heating, water heating, air conditioning, process heat and transport, rather than in final energy demands for specific fuel types. This approach permits the maximum analysis of interfuel substitution at the point of end use. However, the useful energy approach was not practical for all demand sectors or countries because of insufficient data or the nature of national statistics. In these cases the demand was specified in terms of final energy.

Figure 4

IEA ENERGY RD & D STRATEGY PROJECT

OIL PRICE ASSUMPTIONS

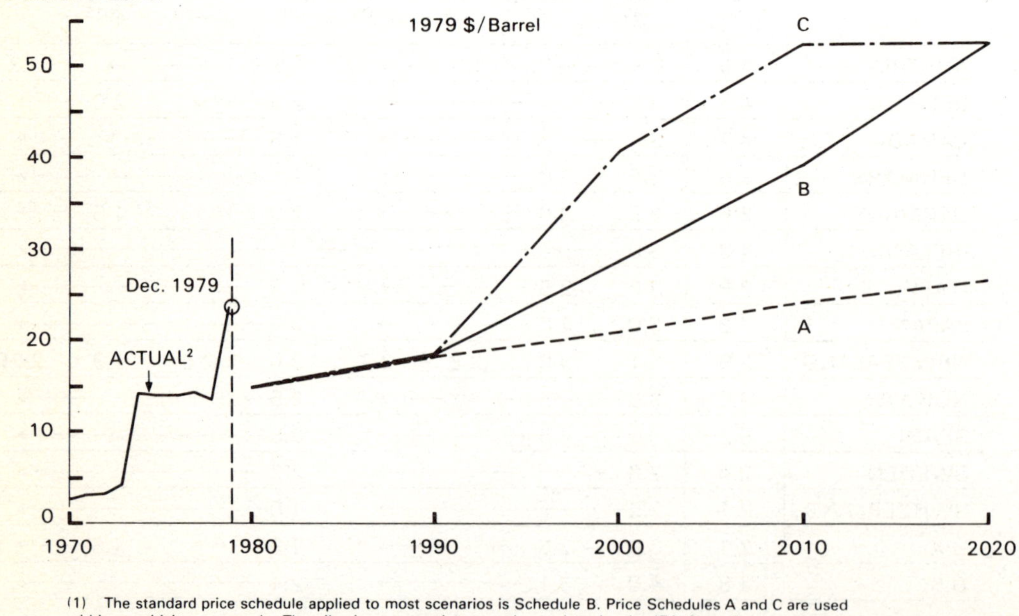

(1) The standard price schedule applied to most scenarios is Schedule B. Price Schedules A and C are used within sensitivity cases only. The oil price assumptions are those given by the IEA Committee on Energy Research and Development (Document IEA/CRD/M(77)47.

(2) Prices for Arabian Marker Crude f.o.b.

The reference set of demands for energy were adjusted for long-range price effects and changes in habits to the greatest extent possible. But the same demands were used for all scenarios, including the alternative oil price cases; that is, no demand elasticity adjustment was made from case to case. In view of the uncertainty of the quantitative effect of energy prices on demand, and the fact that demand estimates in this project only serve as a framework for examining the relative impacts of new technologies, it is not expected that this simplification significantly distorts the results of the project.

The world prices of fuel were established in 1977 on the basis of alternative assumptions as to their future course. The key fuel price is world oil. Three alternative world oil price paths from 1980 to 2020 were examined (Figure 4). Path B is used as the reference price case. Figure 4 also shows actual oil prices for 1973-79 compared to the three alternative future oil price paths. Gradually increasing prices were assumed for other imported fuels and production costs were used for domestic fuel prices. There were slight variations in prices from country to country because of local circumstances. Figure 5 shows a representative set of relative fuel prices for one country.

It is evident that recent oil price movements raise a question over the use of oil price path B as the reference. Even oil price path C appears low. However, a view of the effects of prices higher than path C can be seen in the security scenarios which examine the effects of allowing cost premiums for alternative technologies. (Scenarios SP-1/PREM-1, SP-1/PREM-2, and SP-1/PREM-3 - see section 3 below).

Figure 5

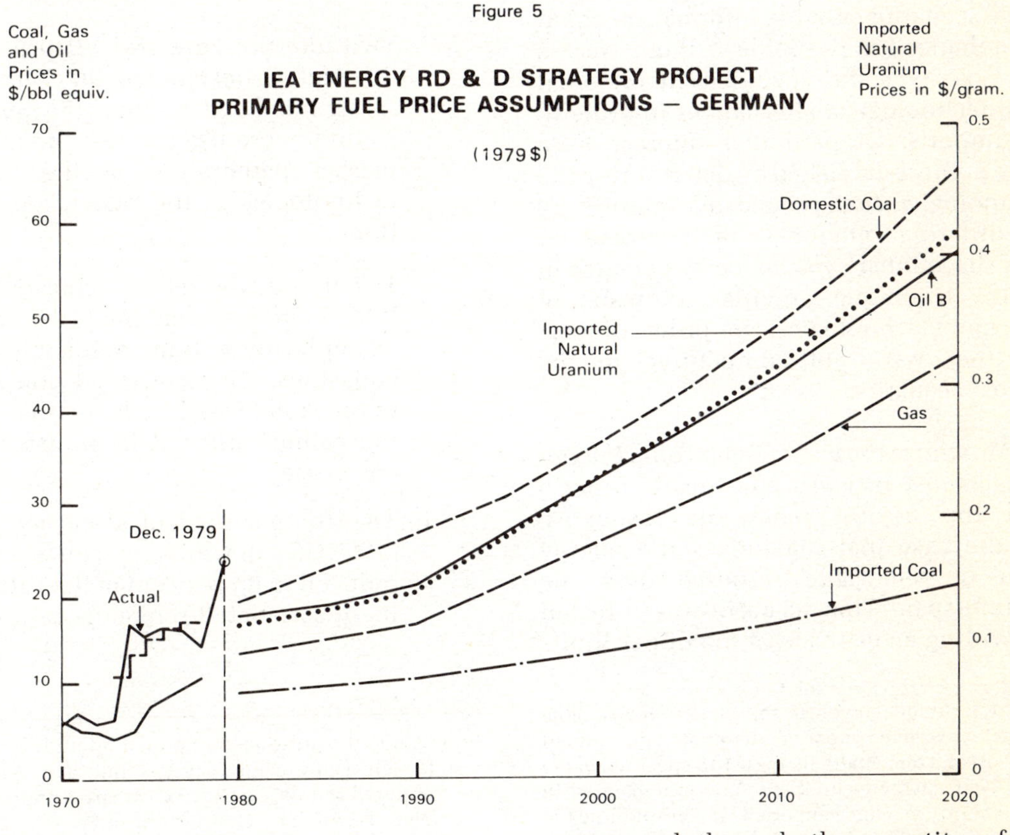

3. The Establishment of Methods and Scenarios for the Analaysis

It was decided to focus the analysis on three key indicators: economic effects represented through the total cost of the energy system (P); security of energy supply represented through the quantity of imported oil (S); and use of renewable energy represented through the quantity of renewable energy contribution (R). The MARKAL Model was designed to analyse energy systems depending upon whether the scenarios emphasized P, S or R. An environmental effect indicator was sought, but no suitable method could be found to reflect in a quantitative manner the diverse environmental effects of the range of technologies. (See Chapter V relating to environmental considerations.)

Scenarios Based on Minimum Costs (designated PS): One set of scenarios gave first priority to minimizing total energy system costs over the forty-year time span and second priority to security of energy supply,

Scenarios Based on Reducing IEA Oil Imports (designated SP): Another set of scenarios gave first priority to security, i.e. reducing oil imports over the forty-year time period, and second priority to minimizing the cost of the energy system. These two sets of scenarios lead to different energy patterns and provide information on technologies desired under either one or both situations.

Scenarios Examining the Trade-Off of System Costs and Security of Supply (designated PREM-1, 2 or 3): Several other scenarios were established to examine intermediate points of trade-off between security of supply and cost of the energy system as a whole. These scenarios provide information about which technologies are desirable if there were a willingness to pay some specified incremental premium for technological alternatives in order to reduce oil imports. Of particular interest were intermediate points established by determining the mix of technologies that reduce oil imports by assuming differing amounts of incremental increases in discounted systems costs per unit of energy.* This technique provides a means of identifying new technologies, in order of cost preference, that were not competitive in the minimum cost scenarios.

Scenario to Maximize the Use of Renewable Energy Sources (designated RP): An additional scenario emphasized the use of renewable resources, defined as the case that maximizes the energy contribution of renewable resource over the forty-year time span. This scenario was directed towards providing an estimate of the upper limits of renewable resource use in the event that energy policy decisions were undertaken to give a maximum push, at whatever cost, to these technologies.

Scenarios Exploring the Effects of Accelerating the Introduction of New Technologies (containing the designation 4): These scenarios explored the effects of accelerating the use of new technologies by advancing the date or increasing the rate of commercial introduction of new technologies. Governments may take action through increased RD & D efforts and/or through incentives and regulation to influence the date or rate of availability of a technology.

Sensitivity Studies: The last focal point of the methodology was to examine the sensitivity of the energy contributions of individual technologies to changes in three major external conditions:

i) *Two alternative world oil price scenarios, one higher (designated Oil C) and one lower (designated Oil A)* than the reference projection, were used to test the effects on the market penetration of the various new technologies in the case of oil price variations.

ii) *Two limited fossil fuel scenarios (designated LIM FOS),* which may be caused, for example, by a limit being placed on CO_2 emissions. The analytical objective was to reduce total fossil fuels use to 65 per cent of the value indicated in an associated reference case.

iii) *Two limited nuclear scenarios (designated LIM NUC),* in which total nuclear energy was reduced to 65 per cent of the value indicated in an associated reference case.

---

* The analytical technique involves the choice of available energy forms that will reduce imported oil until the discounted energy system costs incrementally increase to a specified value (the premium). New energy technologies are implemented in the order of their ability to reduce imported oil while adding the least discounted cost to the system. A higher premium may result either in an earlier date of implementation for a technology or in the inclusion of an energy technology not previously chosen. Present value premiums of \$3.75/barrel, \$7.50/barrel, and \$11.25/barrel (1979 US\$) were investigated. These are identified respectively as PREM-1, PREM-2, and PREM-3. With the use of the 6 per cent discount rate for present value, the premium available for a new technology increases by 6 per cent each year in current year values.

Another way to describe the approach is that the premium can be added to the prescribed imported oil price schedule in each year to yield a new technology cost limit. This relationship may be computed as follows (e.g. \$3.75/barrel premium):

$$(\text{New Technology Cost Limit})_t = (\text{Oil price})_t + \$3.75(1.06)^t$$

where t is number of years beyond 1979

Thus, the approach has the same effects as having investigated higher imported oil price trajectories. However, a new energy technology will not necessarily be implemented even if it has a cost less than the new technology cost limit. Each technology competes against existing and new energy sources for a share of the market to displace imported oil. If that market is captured by one or more existing or new technologies, then a higher cost technology has no market even if its cost is less than the cost limit.

The above studies, in all combinations, could lead to an unmanageable number of scenarios. Criteria were established to select scenarios that appeared to give the greatest information and insights for the development of an RD & D strategy. The scenarios which were retained were those of most interest from the standpoints of energy systems costs, levels of imported oil, their effectiveness in pointing towards desirable futures, and the amount of new information they provided.

A total of 16 scenarios were maintained for the final analysis. These scenarios are identified in Figure 6. Of these two scenarios, one which portrayed the *minimum system cost (PS-1)*, and one which achieved reduced oil imports by the accelerated introduction of new technologies and paying a premium for new technologies *(SP-4/PREM-2) termed accelerated security*, have been used as the principal cases for the discussion and presentation of results.

Figure 6

IEA ENERGY RD & D STRATEGY PROJECT

LIST OF SCENARIOS

Scenario Designation	Scenario Description	New Technology Schedule *
PS–1	Minimum Energy System Cost, Oil B	Non-accelerated
PS–1/Oil A	Minimum Energy System Cost/Lower Oil Price Sensitivity Study	Non-accelerated
PS–1/Oil C	Minimum Energy System Cost/Higher Oil Price Sensitivity Study	Non-accelerated
PS–1 (LIM FOS)	Minimum Energy System Cost/Sensitivity Study on Limited Fossil Availability 65 % of availability in PS–1, Oil B, Scenario	Non-accelerated
PS–1 (LIM NUC)	Minimum Energy System Cost/Sensitivity Study on Limited Nuclear Availability – 65 % of Availability in PS–1, Oil B, Scenario	Non-accelerated
SP–1/PREM-1	Energy System Cost/Security Trade-off (with oil imports reduced by cost premium for new technologies of up to $3.75/bbl equivalent) *	Non-accelerated
SP–1/PREM-2	Energy System Cost/Security Trade-off (with oil imports reduced by cost premium for new technologies of up to $7.50/bbl equivalent) *	Non-accelerated
SP–1/PREM-3	Energy System Cost/Security Trade-off (with oil imports reduced by cost premium for new technologies of up to $11.25/bbl equivalent *	Non-accelerated
SP–1	Minimum Net Oil Import (Maximum Security), Oil B	Non-Accelerated
RP–4	Maximum Use of Renewable Technologies, Oil B	Accelerated
PS–4	Minimum Energy System Cost, Oil B	Accelerated
SP–4/PREM-2	Energy System Cost/Security Trade-off with oil imports reduced by cost premium for new technologies of up to $7.50/bbl equivalent) *	Accelerated
SP–4/PREM-2 (LIM NUC)	Sensitivity Study on Limited Nuclear Availability. Availability is 65 % of Availability in SP–4/PREM-2 Scenario.	Accelerated
SP–4/PREM-2 (LIM FOS)	Sensitivity Study on Limited Fossil Availability. Availability is 65 % of availability in SP–4/PREM-2 Scenario.	Accelerated
SP–4/PREM-2 Oil A	Energy System Cost/Security Trade-off Lower Oil Price Sensitivity Study	Accelerated
SP–4/PREM-2 Oil C	Energy System Cost/Security Trade-off Higher Oil Price Sensitivity Study	Accelerated

* See text for more complete explanation.

4. Limitations of the Approach Employed

The many unknowns and uncertainties inherent in a study of this nature dictate that caution be applied in viewing the numerical results of the systems analysis work as presented in the report. The essence of energy RD & D strategy development has been to gain the insights which the results offer, without attributing to the numbers a precision they do not have.

In addition, the analytical effort, while extensive, nevertheless did not explicitly address a number of issues which have relevance in estimating the future potential of technologies. In its judgements in the development of the recommended strategy, CRD attempted to take these limitations into account; but it is useful to note the most important of them:

- The projected costs of the individual technologies, particularly as they compare with other technologies to supply an energy need, strongly influence the results. While an attempt was made to use the best information available, there are inherent uncertainties in such projections. Sensitivity studies of the effect of changes in estimates of the cost and other characteristics of individual technologies were not performed.
- The maximum rate at which a new technology can be implemented once it becomes commercial has been estimated. The best available information was used in these estimates, although in many cases those estimates are not the result of detailed studies.
- While questions relating to social and environmental impacts are identified, it has not been possible to quantify those impacts.
- It is assumed that trade in energy fuels and technologies among IEA countries is unrestricted. To the extent this assumption proves to be incorrect, the strategy of individual countries and of the IEA group could be affected.
- The analysis is based upon one set of demand projections. Within broad limits it is not expected that variations in demand would affect the relative importance of the technologies studied, with the important exception of major changes in energy use patterns.
- Energy conservation technologies were only partly treated explicitly; much of their effect was included in demand projections. Therefore, the potential contributions of conservation related technologies are understated in the results.

Chapter III

SUMMARY OF SYSTEMS ANALYSIS RESULTS

A. INTRODUCTION

Using the methodology described in Chapter II, fifteen IEA countries each analysed means of satisfying specified energy demands over the time period 1980-2020. This chapter is intended to provide an overview of the key analytical results that influence strategic choices. The results are an aggregation of the fifteen country analyses of all or a portion of the sixteen scenarios. Since it is neither feasible nor desirable to present all of the information in this summary document, for a more detailed examination of analytical results the reader is referred to the forthcoming report of the Steering Group on Energy RD & D Systems Analysis. A summary of these results is presented in Annex III.

B. ENERGY SYSTEM OVERVIEW

Table 3

IEA ENERGY RD & D STRATEGY PROJECT
ENERGY DATA INDICATED FOR 15 COUNTRIES
MINIMUM SYSTEM COST SCENARIO NON-ACCELERATED: (PS-1)

RESOURCE, FUEL TYPE	YEAR (mtoe/year)				
	1980	1990	2000	2010	2020
	PRIMARY ENERGY (FOSSIL FUEL EQUIVALENT)				
COAL*	750	1090	1470	2080	2630
OIL*	1720	1770	1770	1680	1760
GAS	630	760	730	670	550
NUCLEAR	220	470	900	1330	1820
HYDRO	240	310	360	390	400
OTHER RENEWABLES	10	50	140	290	410
TOTAL PRIMARY	3610	4450	5370	6440	7570
	FINAL ENERGY CONSUMPTION				
SOLIDS	210	410	800	1190	1600
LIQUIDS	1460	1600	1630	1580	1700
GASES	610	670	700	720	610
ELECTRICITY	350	480	620	740	870
DISTRICT HEAT	10	20	40	50	50
PROCESS HEAT	10	50	70	120	170
OTHER RENEWABLES (DIRECT USE)	10	20	70	130	180
TOTAL FINAL	2660	3250	3930	4530	5180
	ELECTRICITY PRODUCTION				
FOSSIL	220	240	220	170	120
NUCLEAR	70	160	310	450	620
HYDRO	90	110	140	150	150
OTHER RENEWABLES	0	10	20	50	80
TOTAL ELECTRICITY	380	520	690	820	970

* Coal includes peat and coke imports; oil includes tar sands, oil shale and natural gas condensates.

Table 4

IEA ENERGY RD & D STRATEGY PROJECT
ENERGY DATA INDICATED FOR 15 COUNTRIES
MINIMUM SYSTEM COST SCENARIO: ACCELERATED (PS-4)

RESOURCE, FUEL TYPE	YEAR (mtoe/year)				
	1980	1990	2000	2010	2020
	PRIMARY ENERGY (FOSSIL FUEL EQUIVALENT)				
COAL*	740	1080	1380	2000	2620
OIL*	1710	1710	1680	1540	1630
GAS	670	740	710	630	520
NUCLEAR	220	500	1050	1530	1950
HYDRO	240	310	370	380	390
OTHER RENEWABLES	10	100	210	370	470
TOTAL PRIMARY	3590	4440	5400	6450	7500
	FINAL ENERGY CONSUMPTION				
SOLIDS	210	440	810	1110	1580
LIQUIDS	1370	1540	1550	1500	1600
GASES	610	650	690	690	570
ELECTRICITY	350	460	630	760	890
DISTRICT HEAT	10	20	30	50	50
PROCESS HEAT	10	90	140	190	240
OTHER RENEWABLES (DIRECT USE)	10	30	90	190	240
TOTAL FINAL	2570	3230	3940	4490	5170
	ELECTRICITY PRODUCTION				
FOSSIL	220	220	160	120	100
NUCLEAR	70	170	360	510	660
HYDRO	90	120	140	150	150
OTHER RENEWABLES	0	20	30	50	70
TOTAL ELECTRICITY	380	530	690	830	980

* Coal includes peat and coke imports; oil includes tar sands, oil shale and natural gas condensates.

Energy supply and consumption data for the period 1980-2020 are shown in Tables 3, 4, 5, and 6 for: the minimum cost scenario (PS-1); a scenario based on minimum cost with accelerated effort to introduce new technologies (PS-4); two scenarios to minimize oil imports by paying a premium for new technologies - one without technology acceleration (SP-1/PREM-2) and one with acceleration to further increase security (SP-4/PREM-2). Table 7 indicates the contributions to primary energy from conventional sources in the minimum cost and accelerated security scenarios, and Table 8 gives the percentage of total primary energy supplied by new technologies in the same two cases. The average annual growth rates of primary energy forms in the PS-1 and SP-4/PREM-2 scenarios are shown in Tables 9 and 10. The relative differences in total undiscounted energy system capital investment costs for

Table 5

IEA ENERGY RD & D STRATEGY PROJECT
ENERGY DATA INDICATED FOR 15 COUNTRIES
NON-ACCELERATED SECURITY SCENARIO (SP-1/PREM-2)

RESOURCE, FUEL TYPE	YEAR (mtoe/year) 1980	1990	2000	2010	2020
	PRIMARY ENERGY (FOSSIL FUEL EQUIVALENT)				
COAL*	760	1160	1590	2190	2730
OIL*	1700	1600	1560	1380	1410
GAS	680	860	820	810	730
NUCLEAR	220	480	910	1360	1920
HYDRO	240	310	360	390	410
OTHER RENEWABLES	10	60	180	380	520
TOTAL PRIMARY	3610	4470	5420	6510	7720
	FINAL ENERGY CONSUMPTION				
SOLIDS	210	460	800	1020	1430
LIQUIDS	1460	1460	1480	1440	1500
GASES	620	760	800	870	780
ELECTRICITY	350	480	630	770	940
DISTRICT HEAT	10	20	40	50	50
PROCESS HEAT	10	50	70	120	180
OTHER RENEWABLES (DIRECT USE)	10	30	90	150	230
TOTAL FINAL	2670	3260	3910	4420	5110
	ELECTRICITY PRODUCTION				
FOSSIL	230	240	220	170	130
NUCLEAR	70	160	310	470	660
HYDRO	90	120	140	150	160
OTHER RENEWABLES	–	10	30	70	100
TOTAL ELECTRICITY	390	530	700	860	1050

* Coal includes peat and coke imports; oil includes tar sands, oil shale and natural gas condensates.

Table 6

IEA ENERGY RD & D STRATEGY PROJECT
ENERGY DATA INDICATED FOR 15 COUNTRIES
ACCELERATED SECURITY SCENARIO (SP-4/PREM-2)

RESOURCE, FUEL TYPE	YEAR (mtoe/year) 1980	1990	2000	2010	2020
	PRIMARY ENERGY (FOSSIL FUEL EQUIVALENT)				
COAL*	750	1110	1590	2230	2790
OIL*	1680	1560	1460	1260	1240
GAS	680	860	760	730	700
NUCLEAR	220	520	1060	1590	2090
HYDRO	240	310	380	390	400
OTHER RENEWABLES	10	100	270	420	570
TOTAL PRIMARY	3580	4460	5520	6620	7790
	FINAL ENERGY CONSUMPTION				
SOLIDS	220	440	820	1040	1400
LIQUIDS	1440	1430	1410	1350	1390
GASES	610	760	750	790	770
ELECTRICITY	350	480	650	800	970
DISTRICT HEAT	10	20	30	40	50
PROCESS HEAT	10	90	140	190	240
OTHER RENEWABLES (DIRECT USE)	10	40	160	250	330
TOTAL FINAL	2650	3260	3960	4460	5150
	ELECTRICITY PRODUCTION				
FOSSIL	220	220	180	200	140
NUCLEAR	70	170	370	540	710
HYDRO	90	110	140	150	130
OTHER RENEWABLES	0	20	30	60	80
TOTAL ELECTRICITY	380	520	720	950	1060

* Coal includes peat and coke imports; oil includes tar sands, oil shale and natural gas condensates.

four scenarios are shown in Figure 7. Figure 8 contains five curves derived from the accelerated security case showing the sources of gas, electricity, liquids and heat (in the residential and commercial sectors) and the uses of liquid fuels.

The following review of the relatives results of the minimum cost case (PS-1) and the accelerated security case (SP-4/PREM-2) portrays the differing effects of these two policy thrusts.*

*The discussion that follows does not generally treat results past 2010 due to the greater uncertainty of such numbers, but they are used to identify important trends.

Table 7

IEA ENERGY RD & D STRATEGY PROJECT
CONVENTIONAL SOURCES OF PRIMARY ENERGY
mtoe

NON-ACCELERATED MINIMUM ENERGY SYSTEM COST SCENARIO, OIL PRICE B (PS-1)

	1980	1990	2000	2010	2020	AVERAGE 1990-2020
COAL	750	1090	1470	2070	2620	1810
OIL	1710	1610	1470	1300	1250	1410
GAS	630	760	640	500	360	570
NUCLEAR	220	450	810	1030	1110	850
HYDRO	240	310	360	370	370	350
OTHER RENEWABLES	10	20	40	100	150	80
TOTAL PRIMARY	3560	4240	4790	5370	5860	5070

ACCELERATED SECURITY
SCENARIO, OIL PRICE B
(SP-4/PREM-2)

	1980	1990	2000	2010	2020	AVERAGE 1990-2020
COAL	750	1110	1590	2220	2780	1930
OIL	1650	1410	1090	830	770	1030
GAS	680	790	670	560	480	630
NUCLEAR	220	480	920	1160	1210	940
HYDRO	240	310	360	370	380	360
OTHER RENEWABLES	10	50	80	120	160	100
TOTAL PRIMARY	3550	4150	4710	5260	5780	4990

Table 8

IEA ENERGY RD & D STRATEGY PROJECT NEW SOURCES OF PRIMARY ENERGY PERCENTAGE CONTRIBUTION TO TOTAL PRIMARY ENERGY SUPPLY

MINIMUM ENERGY SYSTEM COST
SCENARIO NON-ACCELERATED,OIL PRICE B
(PS-1)

	1980	1990	2000	2010	2020	AVERAGE 1990-2020
COAL	0	0	0.1	0.1	0.1	0.1
OIL	0.5	3.7	5.6	6.0	6.7	5.7
GAS	0	0	1.7	2.6	2.5	1.9
NUCLEAR	0	0.4	1.6	4.7	9.4	4.7
HYDRO	0	0.1	0.1	0.3	0.3	0.2
OTHER RENEWABLES	0	0.6	1.9	3.0	3.4	2.4
TOTAL PRIMARY	0.5	4.8	11.0	16.7	22.4	15.0

ACCELERATED SECURITY
SCENARIO, OIL PRICE B
(SP-4/PREM-2)

	1980	1990	2000	2010	2020	AVERAGE 1990-2020
COAL	0	0	0	0	0	0
OIL	0.9	3.2	6.7	6.4	6.1	5.8
GAS	0	1.4	1.6	2.6	2.9	2.2
NUCLEAR	0.1	0.9	2.7	6.6	11.3	6.2
HYDRO	0	0.1	0.3	0.3	0.3	0.2
OTHER RENEWABLES	0	1.1	3.3	4.5	5.3	3.8
TOTAL PRIMARY	1.0	6.7	14.6	20.4	26.0	18.2

Table 9

IEA ENERGY RD & D STRATEGY PROJECT
RELATIVE USE AND AVERAGE ANNUAL GROWTH RATE OF PRIMARY ENERGY
MINIMUM SYSTEM COST SCENARIO (PS-1)

	1980 % of Total	Av. Annual Growth %	1990 % of Total	Av. Annual Growth %	2000 % of Total	Av. Annual Growth %	2010 % of Total	Av. Annual Growth %	2020 % of Total
COAL	21	3.8	24	3.0	27	3.0	32	3.5	35
OIL	48	0.3	40	0.0	33	0.5	26	0.5	23
(OIL IMPORTS)	(31)	(-2.9)	(23)	(-1.9)	(19)	(-3.0)	(14)	(-2.4)	(11)
GAS	19	1.2	17	0.4	14	0.8	10	2.0	7
NUCLEAR	6	7.9	11	6.7	17	4.0	21	3.2	24
HYDRO	7	2.6	7	1.5	7	0.8	6	0.3	5
OTHER RENEWABLES		17.5	1	10.8	3	7.6	5	3.5	5
TOTAL PRIMARY	100	2.1	100	1.9	100	1.8	100	1.6	100

For Absolute Numbers see Table 3.

Table 10

IEA ENERGY RD & D STRATEGY PROJECT
RELATIVE USE AND AVERAGE ANNUAL GROWTH RATE OF PRIMARY ENERGY
ACCELERATED SECURITY SCENARIO (SP-4/PREM-2)

	1980 % of Total	Av. Annual Growth %	1990 % of Total	Av. Annual Growth %	2000 % of Total	Av. Annual Growth %	2010 % of Total	Av. Annual Growth %	2020 % of Total
COAL	21	4.0	25	3.7	29	3.4	34	2.3	36
OIL	47	0.7	35	0.7	26	1.5	19	-0.2	16
(OIL IMPORTS)	(30)	(-4.5)	(19)	(-4.5)	(12)	(-5.3)	(7)	(-3.3)	(5)
GAS	19	2.4	19	1.2	14	0.4	11	0.4	9
NUCLEAR	6	9.0	12	7.4	19	4.1	24	2.8	27
HYDRO	7	2.6	7	2.1	7	0.3	6	0.3	5
OTHER RENEWABLES		25.9	2	10.4	5	4.5	6	3.1	7
TOTAL PRIMARY	100	2.2	100	2.2	100	1.8	100	1.6	100

For Absolute Numbers see Table 6.

Figure 7

AVERAGE ANNUAL UNDISCOUNTED CAPITAL INVESTMENT INDEX FOR 15 COUNTRIES UNDER FOUR SCENARIOS

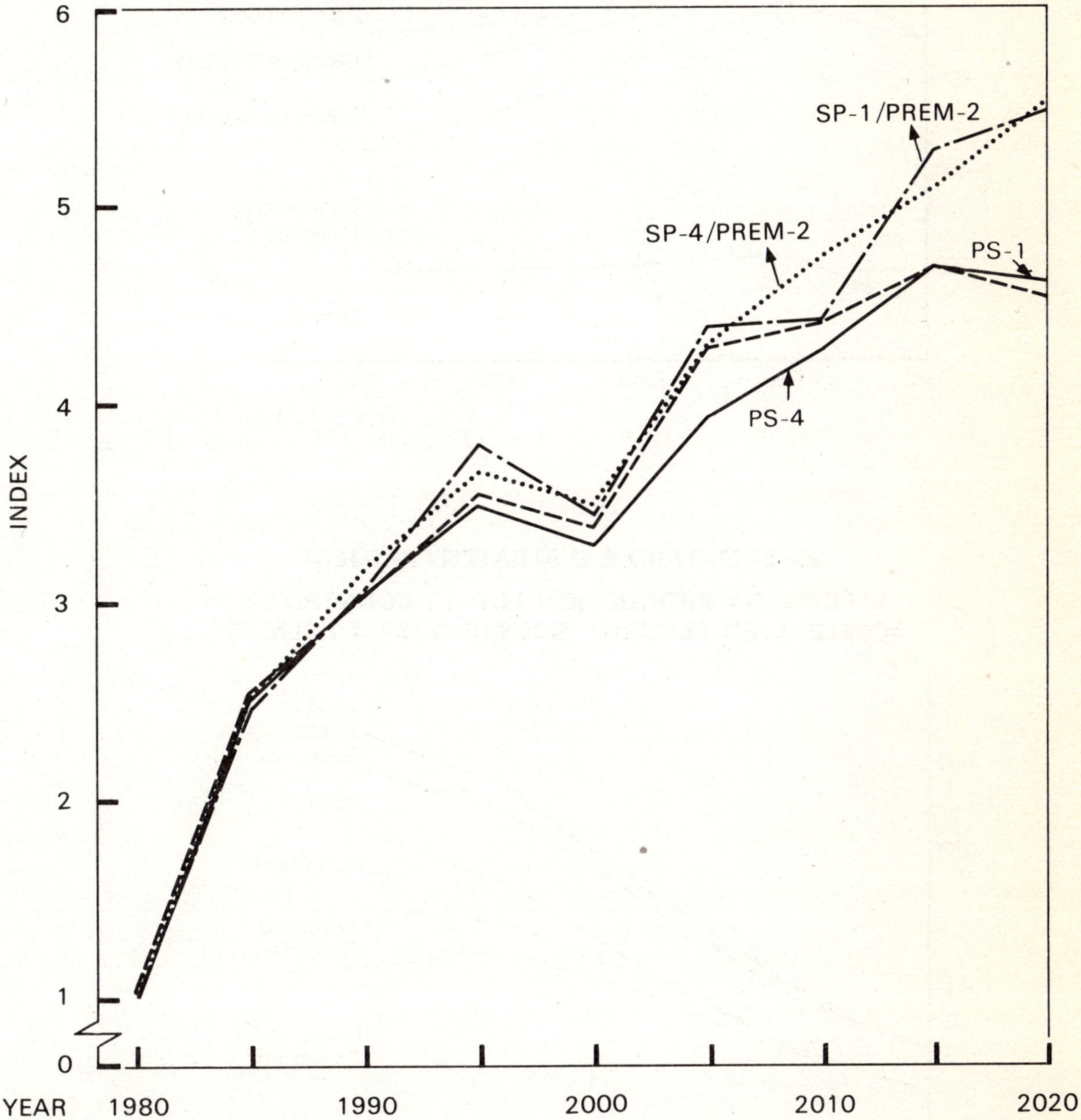

Figure 8 (a)

IEA ENERGY RD & D STRATEGY PROJECT

SOURCES OF GAS FOR 15 COUNTRIES : ACCELERATED SECURITY SCENARIO (SP-4/PREM-2)

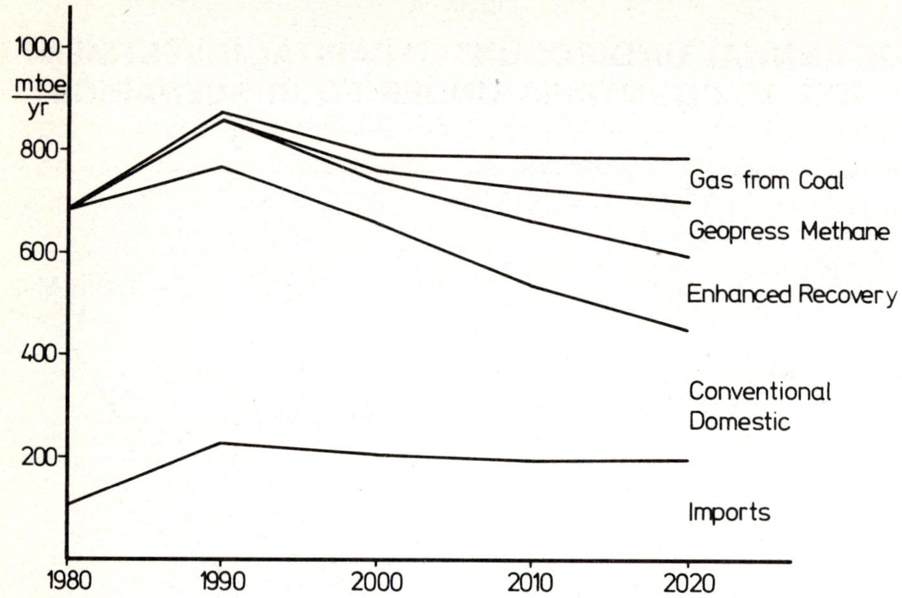

Figure 8 (b)

IEA ENERGY RD & D STRATEGY PROJECT

ELECTRICITY PRODUCTION FOR 15 COUNTRIES : ACCELERATED SECURITY SCENARIO (SP-4/PREM-2)

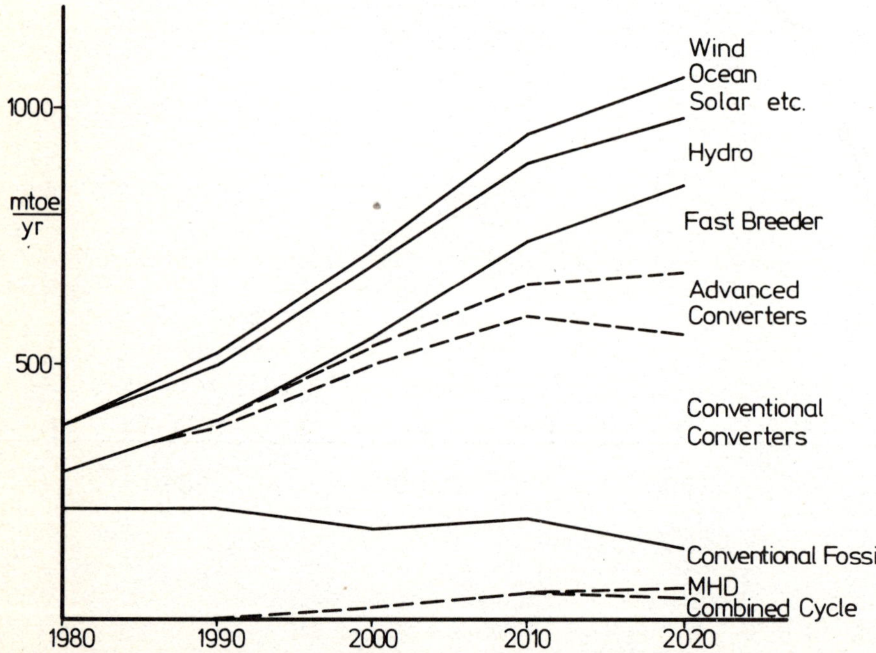

Figure 8 (c)

IEA ENERGY RD & D STRATEGY PROJECT

SOURCES OF LIQUID FUELS FOR 15 COUNTRIES : ACCELERATED SECURITY SCENARIO (SP-4/PREM-2)

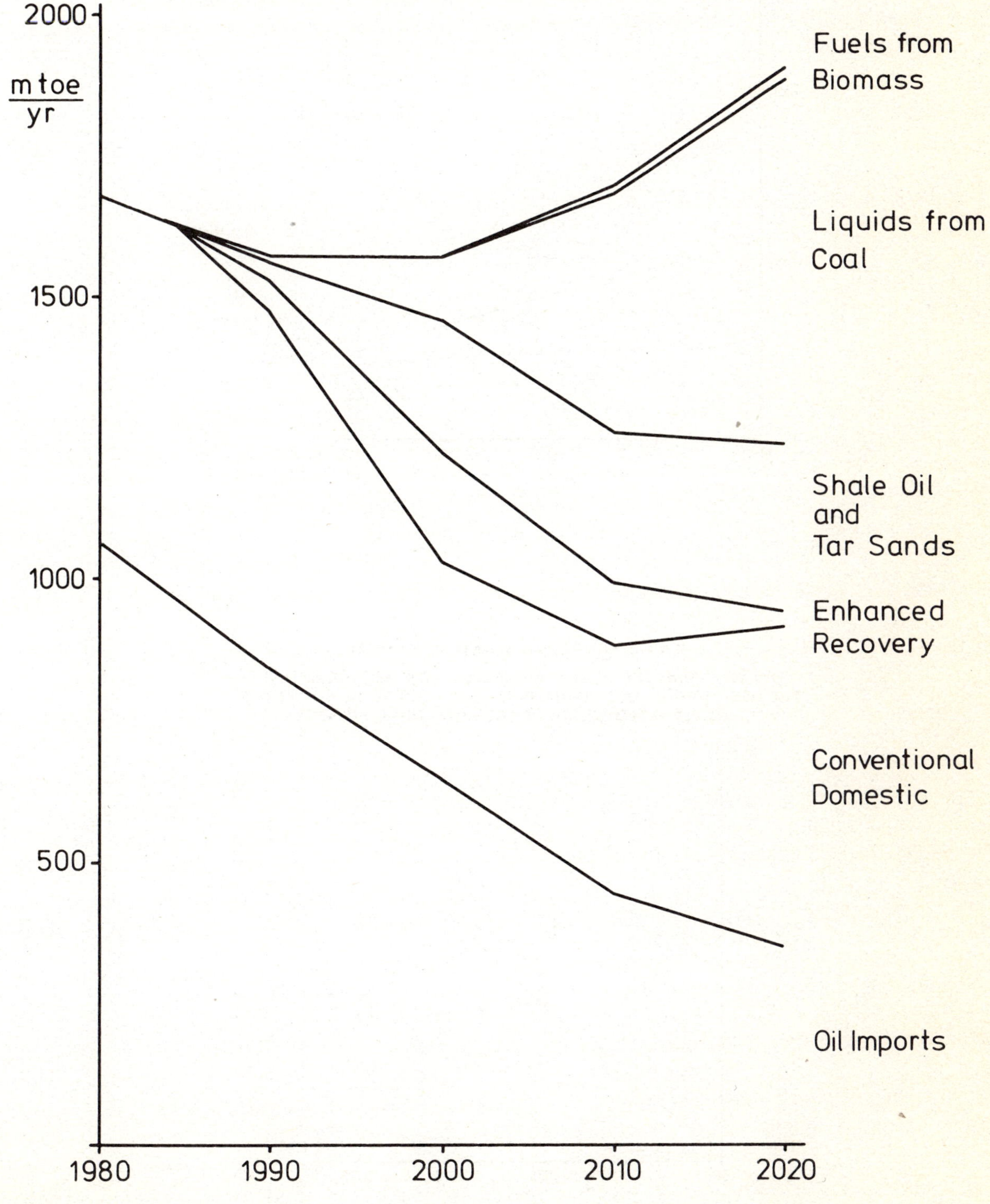

Figure 8 (d)

IEA ENERGY RD & D STRATEGY PROJECT

LIQUID FUELS USE IN 15 COUNTRIES: ACCELERATED SECURITY SCENARIO (SP-4/PREM-2)

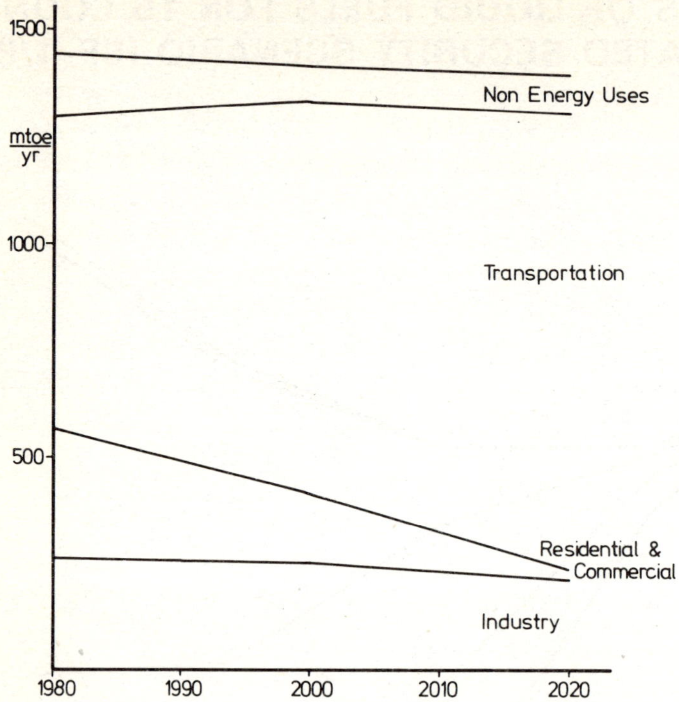

Figure 8 (e)

IEA ENERGY RD & D STRATEGY PROJECT

LOW TEMPERATURE HEAT SUPPLIED BY NEW TECHNOLOGIES IN THE RESIDENTIAL AND COMMERCIAL SECTORS IN 15 COUNTRIES: ACCELERATED SECURITY SCENARIO (SP-4/PREM-2)

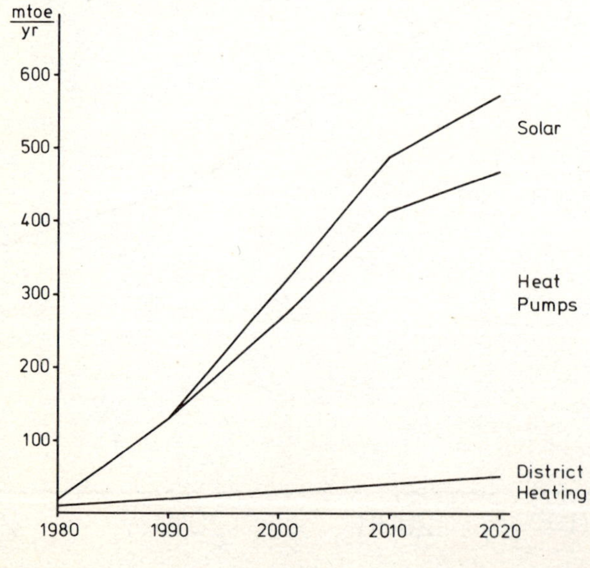

1. Oil and Gas

General. In 1980, oil is estimated to provide about 48 per cent and gas about 19 per cent of the total primary energy supply for the participating IEA countries. In the minimum cost (PS-1) scenario, the combined factors of increased world oil prices (imported gas prices are linked to oil prices) and the availability of new technology project a major shift away from these sources in relative terms. As a per cent of total supply, oil declines to 26 per cent and gas declines to 10 per cent of total primary energy supply by 2010.

In terms of absolute quantities, projected conventional oil requirements remain rather stable in this case, ranging from 1770 to 1670 Mtoe.* Imported oil shows a continuous decline from approximately 1130 Mtoe in 1980 to 860 Mtoe in 2010. The use of natural gas remains stable over time in the minimum cost scenario. The requirements are about 670 Mtoe in 1980, rising to a peak of about 730 Mtoe in 2000, and declining to 670 Mtoe in 2010.

In the accelerated security scenario (SP-4/PREM-2), the projected shift in oil is more dramatic, falling to 19 per cent of supply by 2010. Gas accounts for 11 per cent of supply in this case. The earlier availability in greater quantity of new technologies leads to a much larger reduction of conventional oil use. The oil requirements drop to 1460 Mtoe by 2000 and to 1260 Mtoe by 2010. Oil imports drop to 660 Mtoe in 2000 and to 430 Mtoe in 2010. The peak natural gas use occurs at about 850 Mtoe in 1990, declining to 730 Mtoe in 2010.

*1 million barrels per day = approx. 50 Mtoe per year.

New Technologies. Four types of advanced oil and gas technologies were investigated. (Technologies for transforming coal to oil and gas are discussed in the coal section.) The implementation of enhanced oil recovery and oil shale/tar sands technologies does not appear likely to occur outside the North American countries. Both enhanced oil recovery and tar sands become competitive in the early 1980's while oil shale became competitive in the mid-1990's in the minimum cost case. In both the higher oil price case and the accelerated security case, oil shale became competitive several years earlier. Because the product of enhanced oil recovery is used as early as possible and in large quantities, the results indicate a rise to about 130 Mtoe by 2000 and then a decline due to limited availability of the resource. The contributions from oil shale and tar sands continue to increase rapidly in the post-2000 period.

The application of enhanced gas and geopressurized methane recovery techniques also appears to be limited to North American countries. Both of these technologies become competitive in the mid-to-late 1990's. The enhanced gas recovery shows earlier success in the accelerated technology scenarios. The combined categories project energy contributions of 90 Mtoe in 2000 and 170 Mtoe by 2010.

2. Conservation

General. While this project focussed primarily on new conservation technologies,** the potential of

** Existing technology is defined as that not requiring further RD & D; however, this does not imply that the technology has yet been deployed in all countries.

Table 11

Impacts of New Conservation Technologies in the Minimum System Cost Scenario

	1990	2000 (Mtoe/year)	2010
Energy Efficiency			
Building equipment (mainly heat pumps)	40	90	170
Industrial conservation	60	115	165
Automotive transportation systems	30	170	370
End-Use Substitution			
Electric auto	0	2	5
Alternative Transportation Fuels	20	2	60

existing conservation technologies is large and must not be overlooked. As noted earlier, it was not possible for all countries to treat conservation technologies uniformly. A number of countries included the effects of several types of conservation technologies in their projections of energy demand. Consequently, the results presented here only include the technologies treated directly through the modelling process, and, therefore, most certainly understate the true potential of applying new conservation technologies for the aggregate of IEA countries. The values used here (and given in Tables 16 and 17) are comparable to energy supply values, meaning that additional primary energy input would have been required in the absence of these technologies.

The combination of existing and new conservation technologies showed the potential of contributing 370 Mtoe in 1990, 740 Mtoe in 2000, and 1120 Mtoe by 2010 in the minimum cost case.* In the accelerated security scenario, the contributions rose by 200 Mtoe in 2010. Furthermore, implementation of the new conservation technologies began occurring in significant quantities in the earliest years of the analysis.

* The energy values designated as contributions from conservation technologies are the sum of two components, the fuel input into the device and the end-use energy savings relative to a conventional device, in order to make the numbers comparable to supply technologies. This convention was adopted because the emphasis on evaluating new conservation technologies was to obtain a measure of the extent of their utilisation, not simply to measure reductions in energy requirements. Therefore, the energy values for conservation technologies *cannot* be viewed in the traditional sense as only savings in primary energy inputs.

New Technologies. It is useful to divide the conservation technologies into two groups, those which emphasize energy efficiency and those which represent substitution of other end-use devices for those which use oil. The types of new conservation technologies and the magnitudes of their impact in the reference minimum cost case are shown in Table 11.

As these numbers indicate, the conservation technologies become attractive very early and in significant quantities. These technologies were used by all countries which were able to incorporate them into the model, with the exception of the electric auto, which shows the least promise of the conservation technologies.

3. Coal

General. In the minimum cost (PS-1) scenario, the consumption of coal is projected to increase from 750 Mtoe in 1980 to 2080 Mtoe in 2010. This represents an increase from 20 per cent in 1980 to 32 per cent of total energy supply in 2010. In the accelerated security case (SP-4/PREM-2) the consumption of coal is further increased to 2230 Mtoe in 2010, or 34 per cent of total supply.

The utilization of coal in the PS-1 and SP-4/PREM-2 scenarios in 2010 as compared to that in 1980 is shown below; and the development of coal imports is shown on the table following:

Table 12

Coal Utilization in Two Scenarios

(Mtoe/year)

		Minimum Cost (PS-1)	**Accelerated Security (SP-4/PREM-2)**
Sector	**1980**	**2010**	**2010**
Electricity and heat	450	480	470
Direct Industry	75	935	790
Direct Residential and Commercial	20	5	2
Liquefaction	0	380	595
Gasification	10	110	100
Transport	2	0	0
Coke ovens	155	215	215
TOTAL (Mtoe/year)*	712	2045	2172
TOTAL (Mtce/year)*	1015	2920	3100

* Excludes peat and imported coke.

Table 13

Total Net Coal Import Requirements for 15 IEA Countries Under Two Scenarios (Mtoe/year)

	1980	1990	2000	2010	2020
Minimum Cost (PS-1)	70	100	190	300	500
Accelerated Security (SP-4/PREM-2)	80	130	250	440	640

The increases in coal use introduce three areas of concern, namely, the potential CO_2 problem, effluents other than CO_2, and the practicability of mining and transporting these quantities of coal. The first concern is being externally examined and requires urgent effort to assess its effects. The second can be limited, at some cost, through existing environmental control systems but will ultimately require improved control technologies. Due to the retirement of existing facilities which have less control capability, and with improved control technologies, the total SOx, NOx and total particulate emissions are not expected to increase in proportion to the threefold increase in coal consumption.

With respect to the third area of concern, while the analysis indicates that the expansion of coal can be economically attractive, factors exist which could significantly impact price and availability during the expansion period, (e.g. environmental, industrial and trade policies, and availability of manpower). The IEA report on Steam Coal Prospects to 2000 published in 1978 and the subsequent establishment of IEA coal policies address these issues. In addition, this degree of expansion could lead to greater pressure on coal prices than has been assumed. However, the effects of somewhat higher prices do not appear to alter the relative usefulness of coal technologies.*

New Technologies. Eight new coal technology categories were investigated. These categories included several types of processes for transforming coal to either liquids or gas, new processes for generating electricity, and improved direct combustion of coal for industry.

The principal use of coal in new technology applications was coal liquefaction. However, in the minimum cost case, coal liquefaction did not become competitive until 2000. (Some quantities of coal liquefaction were included in early years in accordance with government-sponsored projects.) Its energy contribution greatly increased in the higher oil price case, and the accelerated-security case where coal liquefaction became competitive by 1990 and was used at some point in every country. The energy contribution became very large in the post-2000 time period in both scenarios.

Pipeline quality high caloric gasification of coal was the next most used new coal technology. The technology was employed in six countries with the dates of introduction ranging from 1985 to 2020. Several countries had sufficient quantities of natural gas, or access to natural gas imports, to make coal gasification technology unnecessary. Industrial quality (low-medium caloric) gasification was attractive to only two countries in small quantities beyond 2000. However, this technology was not modelled for process heat applications or non-fuel uses in several countries and may not have been properly investigated for its full potential.

Of the new coal electric generating technologies, the combined cycle, either with gasification or pressurized fluidized bed, showed greatest use, with substantial impact in two countries and modest impact in five to seven other countries, depending on the case.

MHD, coal gas fuel cells, and in-situ coal gasification technologies were rarely competitive and only then in the very late time periods of the analysis.

4. Nuclear Power

General. The expansion of nuclear power from about 6 per cent of total energy supply in 1980 to about 21 and 24 per cent in 2010 respectively in the minimum cost and accelerated security cases is a key element in the supply system. The magnitude of installed capacity resulting from the analysis is about 575 GWe (310 Mtoe)** in 2000 and 850 GWe (450 Mtoe)** in 2010 (PS-1). In view of the import-

*A higher price for coal was used in test cases in three countries. The price of coal was escalated at a 3 per cent per annum real increase, beginning in 1985, as contrasted with an approximately 2 per cent per annum real increase in its base price schedule. Except for increasing energy system costs, very little change was observed in the new technology contributions.

**Annual Output.

ance of the projected nuclear power contribution, the current uncertainties about the future expansion of nuclear power cause great concern.

Some of the nuclear reactor RD & D programmes are aimed at the use of nuclear power for producing heat. However, most advanced nuclear power development has the purpose of extending the energy availability from naturally occurring uranium and thorium. Thus, in the event that nuclear energy plays the growing role that is called for by the IEA, the analysis indicates that advanced technologies, such as the breeder, will become increasingly important. Furthermore, these technologies are means of improving significantly for some countries the security of energy supply by decreasing dependence on imports.

The cumulative natural uranium requirements in tonnes for nuclear reactors between 1980 and 2020 in the PS-1 and SP-4/PREM-2 cases are given below in Table 14, which also shows reserve and resource estimates:

Table 14

Uranium Requirements in Two Scenarios and WOCA Resources (thousand tonnes)

Uranium Requirements (1980-2020)***		Reasonably Assured WOCA Reserves *	Estimated Additional WOCA Resources *	Speculative WOCA Resources **
PS-1	SP-4/PREM-2			
2,816	3,290	2,590	2,450	6,600-14,800

WOCA = World Outside Communist Areas.

* Source: *Uranium Resources, Production and Demand*, OECD Nuclear Energy Agency, December, 1979.

** Source: NEA Estimate.

*** It should be noted that these cumulative uranium requirements assume recycling of fissionable material, and that the reactors existing in the year 2020 will require for their remaining lifetimes an amount of uranium of the same order as that required from 1980 to 2020.

New Technologies. Two generic types of new nuclear power reactors were investigated: advanced converter reactors and the breeder reactor (Liquid Metal Fast Breeder Reactor).

A wide variety of types of advanced converter reactors were represented in the analysis for use in electric power generation, coupled production, and for gasification of coal. These reactors began to enter the energy system in 1990 and continued to expand in contribution throughout the timeframe of the analysis for eight countries. It should be noted that if countries were unable to develop these new reactors to the stage of commercialization, their market share indicated in the analysis would most likely be captured by existing reactors.

The breeder reactor began making a contribution in the 1995-2000 period and was used in nine countries at some point. The analytical results showed substantial growth rates in the post-2000 period. The availability of uranium was assumed to be generally unlimited, thus countries did not find it necessary to abandon existing reactors in favour of breeder reactors for reasons of uranium shortage (other than uranium supply price effects). Instead, the breeder reactor became the economic method for nuclear-generated electricity as a result of the assumed 3 per cent real cost increase per year for uranium fuel.

5. Renewable Energy Sources

General. The potential for increased use of renewable energy resources in the IEA countries is demonstrated by the rather steady rate of growth of about 3.5 per cent per annum from 1980 to 2010 in the minimum cost scenario. The estimated 1980 contribution of about 250 Mtoe, representing 7 per cent of total primary energy input, consists almost entirely of hydroelectric power. In 2010, renewable energy shows a contribution of 680 Mtoe, or about 11 per cent of the primary energy input. In the accelerated secrutity case, the renewable energy contribution is about 810 Mtoe, or 12 per cent of primary energy input in 2010. One might have expected the contribution to be even greater in this scenario; however, several higher cost forms of renewable energy were not competitive with lower cost non-renewable energy forms which were available earlier through the accelerated schedule.

A measure of the probable upper limit of renewable energy contributions was taken in the accelerated renewables scenario, RP-4. In this scenario, renewables were, in effect, used on accelerated schedules regardless of cost, only limited by end-use demand and other technical constraints. (This was accomplished, for computer calculation purposes, by assuming they had zero capital costs. While zero capital costs were used to gain a view of the maximum possible contribution of renewables, the full estimated capital costs were later included to determine the total energy systems costs of such a scenario.) This led to a contribution from renewables of about 910 Mtoe in 2000 and 1410 Mtoe in 2020. These quantities represent about 17-18 per cent of the total primary energy requirements in these periods. The increases in renewable energy in RP-4 over that contributed in the other cases did not become significant until post-1990.

However, in 2010 in the minimum cost case only 120 Mtoe of the 680 Mtoe resulted from wholly new renewable technologies. A similar pattern exists in the accelerated security case. Thus, the predominant expansion of renewable energy continued to be hydroelectric and existing biomass forms.

New Technologies. Eight categories of renewable technologies were investigated, with a large number of processes being included because of the specialized situations in individual countries. These applications included both the centralized and decentralized direct application of solar energy, large and small scale wind systems, ocean power, fuels from biomass, and geothermal energy.

Both the small-scale and large-scale renewable technologies were found among the competing new processes. The largest energy impacts occurred in geothermal energy, but in the early years the application is limited to six countries where geothermal hydrothermal resources exist. In the very late years of the analysis, additional countries had limited contributions from geothermal hot dry rock.

The next major renewable contributor was the application of direct solar energy for the residential and commercial sector and for low temperature industrial process heat. These technologies become competitive in limited applications in the 1980's. They were applied during the next twenty-five years in two-thirds of the countries, particularly in the accelerated-security case.

Fuels from biomass, in new technology applications, had small but widespread application. Existing biomass technology in expanded use dominated the new applications.

Wind power, ocean power, and centralized solar electric made far less contribution. Wind power did penetrate the market fairly early in some countries, but showed almost no growth potential. Ocean power and centralized solar electric came into use in a few countries very late in the analysis timeframe, although they were available for selection at much earlier dates.

6. Assessment of Overall Energy System Results

The analytical results serve as a reasonable planning base for decisions on RD & D priorities, portraying trends and relative energy contributions from differing technologies and options. However, the analysis was not intended to predict future energy balances. Although the complete analysis included, for most countries, sixteen scenarios to try to capture the effects of uncertainty and the likely future situations, there remains a concern that the results may be too optimistic or too pessimistic about one or more energy classes. These concerns usually translate into questions about what energy form or technology would be most affected if a particular magnitude of one energy class were significantly changed. The most serious concerns that have arisen in the assessment of the results are:

1. Can the quantity of imported oil actually be reduced over time as indicated by these results?
2. Can coal production and use be expanded three-fold over 30 years?
3. Can nuclear power production be expanded to the extent the analytical results indicate?
4. Can commercialization barriers be overcome to permit new technologies to achieve the dates and rates of introduction portrayed in the results?

The answers to these questions are beyond the scope of this project. It is to be noted, however, that these questions reflect a concern that the energy contributions from conventional energy sources implied in the results may not in fact be achieved, which would place an even greater need on new technologies if IEA nations are to avoid serious economic consequences.

C. THE ENERGY SYSTEM COST/SECURITY/TECHNOLOGY ACCELERATION TRADE-OFFS

As noted earlier, one of the focal points of the analysis was an examination of the trade-off between energy system costs and energy security. This trade-off was examined both on the assumption of a normally paced development and deployment of new energy technologies, and on the assumption of an accelerated development and deployment. The aggregated trade-offs for 15 countries are shown in Figure 9. An evaluation of these two curves leads to some very interesting observations of potential impacts that new technologies could help achieve. (It is emphasized that the shapes of these curves vary from country to country as may be seen in the trade-off curves for three IEA countries in Figure 10. From these it will be seen that in some of the individual countries the cost associated with reducing oil imports increased more rapidly than is indicated in the aggregated IEA curve.)

Figure 9

IEA ENERGY RD & D STRATEGY PROJECT

SECURITY / SYSTEM COST / TECHNOLOGY ACCELERATION TRADE-OFF CURVES FOR 15 COUNTRIES

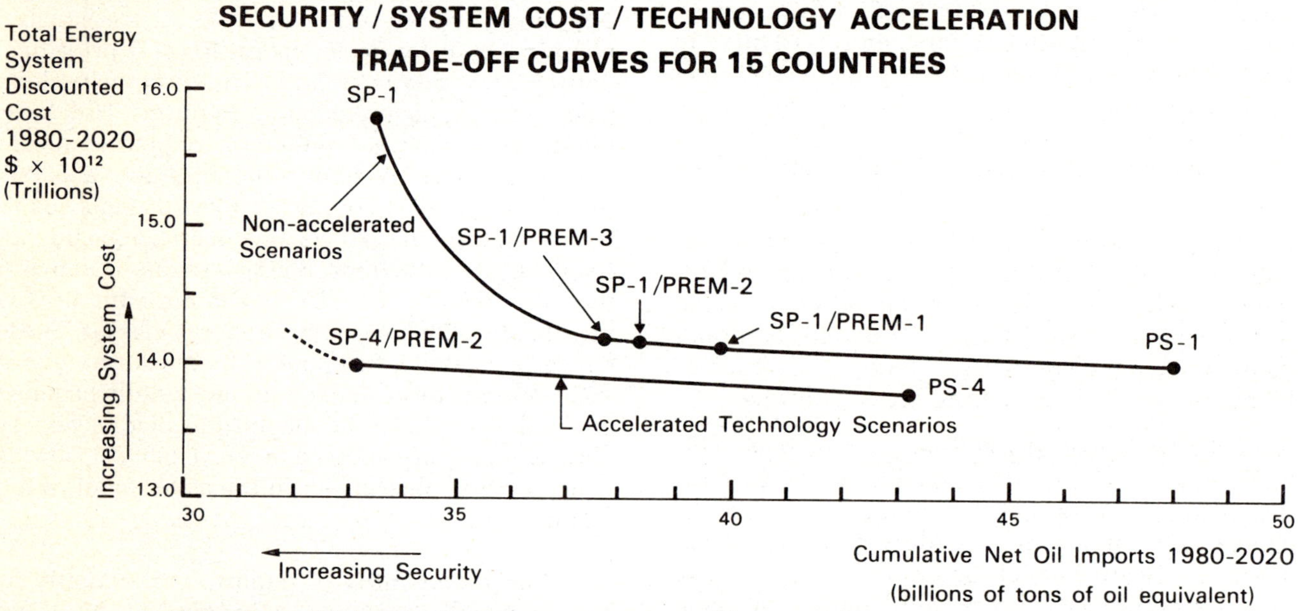

Note: Each point shown on the above curves represents a different scenario. Common to all the scenarios on the non-accelerated cases curve is the assumption that the rate of energy technology development and introduction continues at the present pace. Scenarios on the accelerated technology curve all assume accelerated development and introduction of new technologies.

In Figure 9, the units of measurement are the total energy system discounted costs* over the 1980-2020 period and the cumulative net oil imports over this period. The important quantitative information displayed are the *relative values* and the *change in values* from one scenario to another.

The upper curve assumes current technology development pace (non-accelerated). For the minimum energy system costs (point PS-1), as conventional energy prices rise over time, some new energy technologies enter the energy market. However, at the world oil prices assumed in oil price schedule B, the continued importation of oil in large quantities would appear to cost less than some other alternative technologies.

At the other extreme (SP-1), the emphasis is placed on minimizing oil imports of IEA countries by maximum fuel substitution and introduction of new energy technologies. This analysis indicates that it is conceptually possible to reduce cumulative oil imports by about 30 per cent by incurring for the IEA as a whole approximately a 12 per cent increase in energy system discounted costs ($1.7 trillion). (Again, this percentage increase will vary from country to country.)

---

*The discount rate used in present value calculations was 6 per cent. This value represents a compromise among those countries that are required to use 10 per cent discount rates in evaluating projects and those countries that use a 3-4 per cent discount rate.

With regard to the system cost/security trade-off scenarios, at the first premium for technologies to displace imported oil (SP-1/PREM-1), described earlier, cumulative oil imports would be reduced by about 10 per cent by substituting a few, more expensive technologies for oil consuming ones. The increase in discounted energy system cost is relatively minor in percentage terms (about 0.4 per cent) but is more than $60 billion in present value terms.

Further reductions in oil imports can be achieved by providing higher premiums which permit either earlier introduction of some technologies or the initial use of even higher cost technologies (see SP-1/PREM-2 and SP-1/PREM-3). Increased security of supply becomes more expensive per unit of oil import saved. For a technology development assessment, the identification of the order and timing in which particular groups of new technologies appear along the trade-off curve, provides some measure of the importance of the technology relative to the particular objective.

The lower curve of Figure 11 displays the effects of accelerating the introduction of new technologies. For this examination it was hypothesized that the use of several new technologies could be accelerated by advancing the date of introduction and/or increasing rate of implementation after introduction. These changes from the basic technology characterization data do not assure that a technology will make a greater contribution in the energy system because the technology must still be economically competitive and must still compete with other technologies for the available market share.

The analysis of an accelerated minimum system cost scenario (PS-4) showed that both discounted energy system costs and oil imports could be reduced. This result occurs through increased contributions from direct oil substitutes (enhanced oil recovery, oil shale/tar sands, and coal liquefaction), increased electricity use (advanced converter reactors, breeder reactors and geothermal/hydrothermal), and increased application of all conservation technologies (automotive transport systems, alternative transportation fuels, conservation building equipment, electric auto, and industrial conservation). New technologies that produced gas as the final product showed no change from the PS-1 scenario level. Renewable technologies exhibited a mixed pattern; biomass energy increased, residential and commercial solar heating decreased in the early years and increased in the later years, and the other renewable technologies showed no changes.

One of the most challenging cases is the accelerated security case, shown as the SP-4/PREM-2 point on the accelerated curve. As noted earlier, this scenario incorporates both the acceleration of technologies and the payment of a premium for new technologies up to a specified incremental increase in discounted system costs per unit of energy above the accelerated minimum cost scenario (PS-4). The energy system cost is approximately the same as in the non-accelerated minimum cost scenario (PS-1), excluding additional expenditures to bring about the acceleration. Yet, it would appear to achieve about a 30 per cent reduction in cumulative net oil imports, approximately the same as achieved by the minimum oil import, non-accelerated scenario (SP-1). The cost difference between SP-1 and SP-4/PREM-2, about $1.8 trillion, is an approximation of the savings which accelerated availability of new energy technologies would offer compared to the cost of attempting to reduce imports by 30 per cent by other means.

Again, the identification of technologies that contribute most in the PS-4 and SP-4/PREM-2 scenarios, yields a measure of information around which alternative RD & D strategies can be developed toward alternative objectives.

These figures do not take into account the costs of additional RD & D efforts or the costs of incentives to bring about more rapid introduction. However, the $260 billion reduction in system costs from the PS-1 to PS-4 case is a first order approximation of the additional amounts the IEA countries could expend for that purpose over the next forty years.

Some caution must be expressed at this point. There may be some optimism with regard to the reference dates of availability and rates of new technology implementation in these cases. The results of the accelerated cases should be viewed as an extreme target that could be approached, but not necessarily achieved.

D. MAJOR OBSERVATIONS RESULTING FROM THE ANALYSIS AND THEIR IMPLICATIONS FOR AN IEA RD & D GROUP STRATEGY

1. During the 1980-2020 period, the requirements for imported conventional oil are projected to decline. During the first 20 years, the decline is

mainly caused by conservation and expansion of existing technologies. During the second 20 years, new liquid fuel technologies begin to make a significant contribution.

As shown in Figure 11, the analysis indicates that total oil imports can decline throughout the 1980-2020 period for IEA countries. The analysis should not be interpreted as a prediction that imports will decline, but to indicate that it is feasible to have such a decline.

In the cost minimization case, the indicated oil import decline is rather gradual over time (6 per cent below the 1980 level in 2000, and 18 per cent below in 2010). Earlier studies have indicated that, from a resource standpoint, a decline in the production of world oil can be expected to occur between 1990 and 2010, though it could occur earlier as a result of changes in producer country management of their resources. When the production decline begins, it may be sharp. Therefore, because the decline in oil imports is so gradual, there exists a high risk that its rate of reduction may be insufficient to adjust to the anticipated production decline.

In the accelerated-security case (SP-4/PREM-2), a willingness to pay a premium for alternative technologies could lead to a much greater reduction in demand for oil imports (40 per cent less than the 1980 level by 2000 and 60 per cent less by 2010). If a future pattern approaching the accelerated-security scenario were to develop, there would be an increased probability that world oil production and demand could be kept in balance without further pressure on oil prices.

Strategy Implication: The need to concentrate RD & D efforts on technologies that lead to early and rapid reduction of oil imports is demonstrated if the IEA countries are to diminish the risks of having inadequate availability of supply and increasing price pressure. Because of their important contributions to this effort, the conservation technologies and liquid fuel technologies merit increased attention.

Figure 10

IEA ENERGY RD & D STRATEGY PROJECT

ENERGY SYSTEM COST/SECURITY TRADE-OFF CURVES ILLUSTRATING DIFFERENCES AMONG THREE COUNTRIES (NORMALIZED)

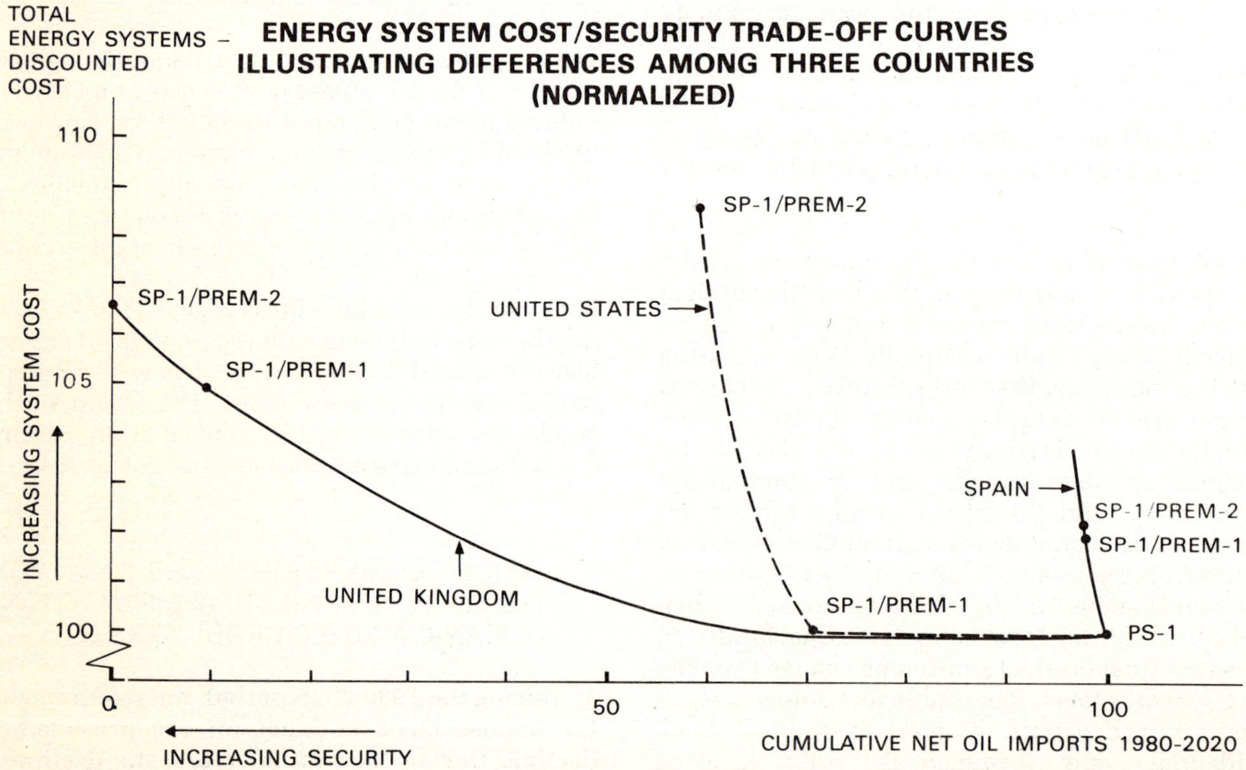

2. The contribution of new energy technologies before the year 2000 may not be adequate unless their introduction is accelerated.

On the basis of oil price assumption B, a number of the new supply technologies are projected not to compete economically before the year 2000. For those new supply technologies which are competitive, the date of introduction is relatively late in the century and the physical constraints on the rate of growth for new technologies combine to limit their aggregate contribution.

Figure 11

IEA ENERGY RD & D STRATEGY PROJECT

NET CONVENTIONAL OIL IMPORTS UNDER FOUR SCENARIOS

(Million tons)

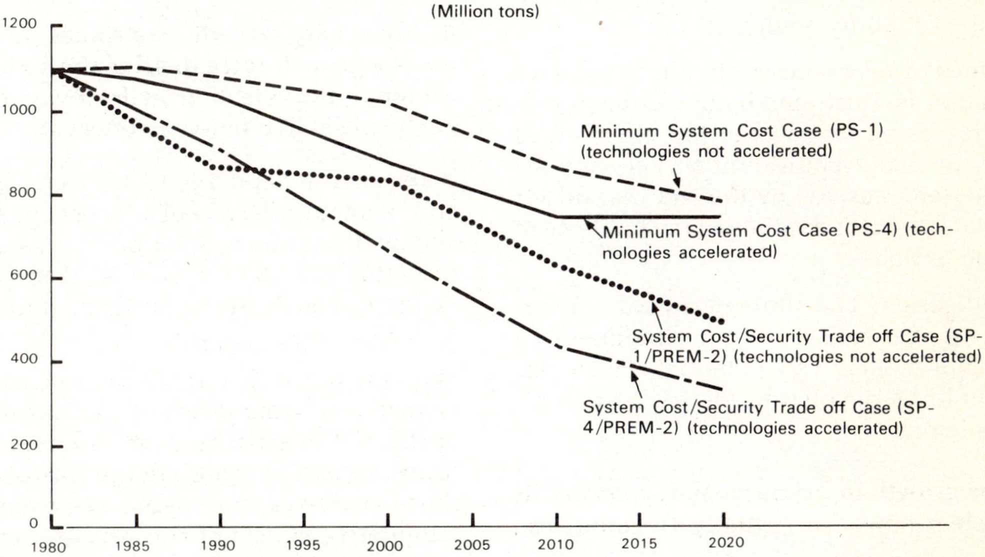

The contribution of new technologies represents about 5 per cent of total primary energy supplies in 1990, about 13 per cent in 2000, and about 23 per cent in 2010 for the minimum cost case. However, in absolute terms, the new supply technologies do not keep pace with increases in primary energy demand until after 2000. One should avoid the conclusion, however, that this contribution from new technologies makes no difference. If new supply technologies were not available, and their replacement fuel were in the form of imported oil, about 750 Mtoe of additional oil imports would be needed in 2000. This would be equivalent to an increase in oil imports by as much as 55 per cent, bringing imports to approximately the same level as 1980.

The advantages of accelerating key technologies and paying some premium cost for early introduction is illustrated in the accelerated-security case. The contribution of new supply technologies increases to 8 per cent of total primary supplies in 1990, to 18 per cent in 2000, and to 30 per cent in 2010. During the period 1990-2000, the increases in new energy technologies are approximately equal to the increase in primary energy demand.

Strategy Implication: It seems mandatory that an accelerated technology path must be pursued. An accelerated path would consist of both increased RD & D efforts, particularly on the most promising of these technologies, and the use of incentives and regulations to assist the market introduction of these technologies. The IEA countries cannot continue on a non-accelerated technology path; the risks are too great and the foregone potential benefits are too large.

3. The impact of new conservation technologies can be very significant in meeting IEA objectives.

Conservation technologies, consisting of more efficient devices, were analysed for all end-use sectors. The utilization of these new technologies, which is entirely separate from conservation due to demand reduction and life-style changes, was very significant, and approximated that of new supply technologies. It would have been greater than the results shown if all countries had been able to make representations of the conservation technologies. However, not all countries could develop sufficient data to permit incorporation of the full range of conservation devices.

In some of the countries, the energy contribution of new conservation technologies exceeds that of new supply technologies.

For the participating IEA countries, new conservation technologies are projected to contribute the equivalent of about 140 Mtoe in 1990 and 400 Mtoe in 2000 in the minimum cost case. The impact is even greater in the accelerated-security case where the contribution is 250 Mtoe in 1990 and 750 Mtoe in 2000. These values are in addition to equally impressive contributions from the expanded use of existing conservation technologies.

The importance of new conservation technologies in transportation is illustrated by three transportation categories being among the top eight of all technologies in energycontribution. The importance is further emphasized by the fact that oil for the transportation sector dominated the total use of oil in all time periods.

Strategy Implication: The strategy should emphasize RD & D on new conservation technologies so they may achieve their full potential. Particular attention should be devoted to the transportation sector technologies.

4. The major growth in primary energy occurs in coal and nuclear power in virtually all countries.

Energy from coal is used in a variety of applications in the scenarios. The most important growth areas for coal use are for coal liquefaction and for direct use by industry, including non-energy uses. In addition, coal continues to be an important contributor o electricity generation, although advanced coal electric technologies did not make large contributions in the period studied.

The largest category of added electricity generation capacity is that of existing nuclear reactors. Beginning around 2000, the breeder contributed very significantly to electricity generation requirements.

Scenarios in which the supply of nuclear power or fossil-based energy are limited suggest that each is capable to some limited degree in making up for shortages in the other. However, a severe problem would probably exist if the growth of either were constrained.

Strategy Implication: Coal development faces its greatest challenges in the ability to produce increasingly large quantities and to be used in an environmentally sound manner. For these reasons, emphasis should be given to mining RD & D and to environmental control technology to ensure that the IEA countries can approach the quantity of coal use indicated by this analysis.

Given the importance of nuclear power in the cases studied, it is clear that the above considerations call for continuing intensive RD & D efforts (1) on nuclear safety and the back-end of the fuel cycle, and (2) on those advanced reactor systems, e.g., the breeder, and other technologies needed to assure an adequate supply of nuclear fuel.

5. Electricity growth continues to exceed total energy growth rates in all of the sixteen scenarios, although the rate of growth slowed down through each successive ten-year interval.

Between 1980 and 1990, electricity growth was at an average annual rate of 3.1 per cent; between 2000 and 2010 the rate was 1.8 per cent, and 1.7 per cent from 2010 to 2020. However, despite the relative decline, electricity remains an important growth factor in energy supply.

The slowing down appears to be caused by five factors: a slowing down of GDP growth; a slowing down of the switching of end-uses to electricity from the use of other energy forms; saturation in those end-uses employing electricity today; the cumulative effects of conservation measures; and, with the exception of heat pumps, the only moderate success of new electric end-use technologies, such as electric vehicles, to penetrate the system. This trend also illustrates the limitations of oil substitution by electricity.

This observation must be tempered somewhat, however, by two considerations: (1) The choices in the model are often made on a system least-cost basis, while historically some electricity using modes, like space heating, have not been made due to least cost but due to convenience and cleanliness. It is very difficult to determine if this situation will continue in the future when energy payments become a much larger portion of consumer budgets. (2) In some applications electricity offers environmental advantages; while an attempt has been made to reflect this element in the costs of alternative sources, the analysis may not fully reflect future trends.

Strategy Implication: Adequate programmes of electricity technologies continue to be warranted. However, the results indicate that the relatively faster rate of growth of electricity may diminish further and that a substantial fraction of energy needs will continue to be met by other forms of energy. This confirms the view expressed in CRD reviews of Member countries' RD & D programmes that RD & D on non-electric energy forms requires greater emphasis.

6. Some new renewable technologies begin to make substantial energy contributions before 2000; however, non-renewable technologies are called upon to provide the major fraction of energy needs even in the case of accelerated efforts to introduce renewables.

A variety of renewable energy technologies were examined with regard to their ability to compete economically in the future and their ability to substitute for oil imports if a premium price were paid. As pointed out earlier, one scenario also examined the impacts of accelerating the date of availability and rate of implementation if the renewables were forced into the system.

In the time interval before 2000, some new renewable technologies are projected to make significant energy contributions in the minimum cost and accelerated-security scenarios. These technologies were: residential and commercial solar heating and cooling, fuels from biomass, and geothermal (hydrothermal).

Most of the new renewable technologies did not begin to be utilized significantly until after 2010. By 2010, the combined contribution of new renewables is 3.4 per cent of the total primary energy in the minimum cost scenarios, 6 per cent of total supply in the accelerated-security scenarios, and 10 per cent of the total energy in the maximum renewables scenario.

Strategy Implication: The analysis suggests that the RD & D effort in new renewables should be primarily focussed on the few most promising technologies, taking into account the fact that there are regional differences in usefulness because of climate and geography, with a lesser, longer term effort on the remaining renewable technologies. This approach should lead to a greater renewables contribution than would be achieved by an RD & D programme which pursued all renewable technologies without such focus.

As pointed out earlier this study was oriented toward the usefulness of new technologies in IEA countries and does not explicitly represent their possible value in other countries. While in many cases the usefulness of new technologies is likely to be similar in other countries, the conditions in many developing countries may in particular offer the potential for greater impact for some of the renewables.

7. The analysis of the consequences of a limitation on the growth in use of fossil energy indicates major effects on economies, energy costs, as well as growing oil and gas imports.

Two sensitivity cases were analysed for each country, in which attempts were made to reduce the quantity of fossil energy cumulatively used over the period by 35 per cent from the quantities resulting from the minimum cost and accelerated security cases. The idea behind these cases was that serious concerns about the global impact of CO_2 and other fossil fuel residuals could possibly lead to such a reduction.

The computation of these cases was performed such that the production and importation of all fossil energy was reduced, without discriminating among certain fuels. However, because of the conversion losses and generally lower end-use efficiencies for coal, reductions in fossil fuels were only observed in coal utilization. Gas virtually retained its market share, while oil imports increased considerably. In addition to the efficiency related switch from coal to oil, the analysis also showed that significant increases in nuclear energy, renewable energy and conservation technologies occurred. The particular technologies showing the greatest change from the associated reference case were the electric auto, biomass, wind power, and nuclear power reactors.

For several countries the attempt to reduce fossil energy by 35 per cent, and still meet the end-use demand, failed. Instead, only 10-20 per cent reductions could be achieved by substituting other technologies for the fossil fuel based ones. These particular cases indicate that further reductions in fossil energy would not be possible without falling short of some demands. A short-fall in energy demand implies an involuntary reduction in economic growth with severe economic perturbations. The total IEA energy systems costs were increased by $1.3 trillion and the oil import quantities were increased by more than 35 per cent in the feasible solutions. These consequences were of greater severity than those resulting from any other single assumption analysed.

Strategy Implication: An aggressive programme of RD & D to prevent this future would focus on environmental control technologies for coal, environmental analysis of CO_2 effects and other global impacts of fossil fuel residuals. Attention in RD & D planning should be given to the substitue and hedge technologies, i.e., nuclear, renewables and conservation, so as to facilitate their increased contribution in the event that a reduction in fossil use becomes necessary.

8. The analysis of the consequences of a limitation on the growth in the use of nuclear energy indicates that increases in not just a few, but in all other forms of energy, would be necessary to offset the partial loss of nuclear energy.

Two sensitivity cases with reduced nuclear contributions were analysed for ten countries (which account for 95 per cent of the projected quantities of nuclear energy for the IEA). In these cases the quantity of nuclear generated electricity, process and district heat was reduced for each country by 35 per cent from the cumulative requirements resulting from the minimum cost and accelerated security cases. The idea behind these cases was to show how delays in nuclear power deployment might affect the supply and technology mix for the IEA. The cases were not designed to study a nuclear moratorium or a particular reduction in the next 20 years. Because of the greater substitution potential in the later years of the period under investigation, the cumulative reductions occurred mainly post-2000, where more new technologies are expected to be available. The analysis indicates that increased oil and gas imports are required in order to offset the partial loss of nuclear power.

The analysis also indicates that IEA's dependence on imported oil and gas continued to be larger than in other cases, thus increasing IEA's vulnerability to supply disruptions. These increases and the cost increases were quite large for some countries.

If a large enough amount and wide variety of new technologies could be made available at some higher costs, this would contribute to offset a part of the nuclear power loss. Increased quantities of central station electricity generation came from higher cost technologies using coal, hydroelectric and renewable sources. Some substitutions in end-use from electric to non-electric forms occurred and coal was diverted from liquefaction processes to electricity generation.

Strategy Implication: The maintenance of growing nuclear power deployment is essential to meet IEA energy demands and IEA objectives for reduced oil imports. RD & D efforts on nuclear safety and technologies for the back-end of the nuclear fuel cycle are necessary. In the event of partially reduced nuclear contributions, the IEA countries would also need to have a large amount and a wide variety of more expensive new technologies, including hedge technologies, available to contribute to the necessary substitutions.

E. EFFECTS OF UNCERTAINTIES

As has been noted, assumptions used in the analysis significantly shape the results of the analyses performed. Thus, it is appropriate to provide some evaluation of the possible effects of the important uncertainties, though it must be recognized that an in-depth examination has not been posible due to time and cost considerations. The most critical assumptions are the world prices of oil, the prices and costs of domestic fuels, the availability and producibility of fuel resources, the economic and associated energy growth rates, and the relative costs of new technologies.

1. Fuel Prices and Availabilities

As pointed out earlier, several differing assumptions of future oil prices were made. A comparison of the scenario results indicates that for oil price B and higher, virtually no change in the date of entry or the energy contribution of the new technologies takes place. This occurs because: (1) technologies that were selected in oil price B scenarios were implemented to their maximum estimated (and allowed) rate of build-up, so that higher oil prices have no effect; and (2) technologies not selected in oil price B cases appear to have costs that continue to exceed even the higher oil prices, or have dates of technical availability that are late in the period covered by the analysis, so that there is very little increase in their contribution. The technologies which do show some increases are coal liquefaction, residential and commercial solar, wind energy, and alternative transportation fuels.

Because each country independently optimized its own scenarios, there existed the possibility that the aggregate oil import requirements could exceed values commonly thought to be available to IEA countries over the long term. Furthermore, unlike other primary sources of energy, experience clearly shows that the amount of world-wide oil production is not only related to demand. For this reason, as well as to enhance security of supply, the IEA has made reductions in oil use and in oil imports central energy policy objectives. The relative amounts of oil which the two principal scenarios call for are shown in Table 15.

Table 15

Oil Imports as a Percentage of Total Oil Requirements in Two Scenarios

	1980	1990	2000	2010
Minimum Cost (PS-1)	66%	61%	57%	51%
Accelerated Security (SP-4/PREM-2)	65%	58%	45%	34%

Thus, the IEA might still find itself importing close to 60 per cent or more of its oil requirements by the end of the century. Retaining such a high level of vulnerability over the next two decades is not compatible with early actions to improve security of energy supply. Only in the accelerated technology security trade-off scenario do important reductions in oil imports appear possible.

The assumptions for prices and costs of fuels other than oil are closely related to the assumptions on availability and producibility of those fuel resources. As described earlier, the analytical results indicate a rather significant increase in coal utilization. In test cases, where a higher set of coal costs were used (see footnote on page 37).

The availability and price of uranium to support the large growth in conventional nuclear power are reasonably aligned with other national and international studies (Table 14). Probably the key concern in this area is the ability to develop and produce annual uranium quantities that keep pace with the indicated number of reactors. Additional exploration to convert speculative resources into uranium reserves will be necessary to provide confidence in future availability and to reduce spot market price pressures.

2. Economic Growth

The analysis did not examine the effects of different economic growth rates on energy futures and the new technology prospects. However, several national studies on this issue have generally reached the conclusion that, while higher or lower economic growth rates resulted in higher or lower energy growth, the *relative* timing, impacts, and priorities of new technologies remained essentially unchanged. The major variation in the conclusion is that with higher growth, conservation technologies acquire even greater importance.

3. Technology Characterizations

The performance of new technologies in the analysis, and consequently the conclusions of the RD & D strategy, can be significantly affected by the cost and efficiency parameters developed in the technology characterization process. The most frequently asked question is how would the strategy, in general, and technology priorities, in particular, be altered if one or more technology cost representation was changed. Since it has not been possible to establish, document, and make the many computer runs to examine in detail various cost estimates for each technology, evaluation of the effect of technology costs must be handled by a combination of expert review of the basic data, observation of technology performance in a few sensitivity cases, use of certain marginal cost information and judgement.

Experience has shown that the costs of technologies tend to be under-estimated, particularly when still in early developmental stages. As a result, new technologies might not become competitive as early as estimated. Such an effect would require greater use of conventional sources and potentially greater reliance on imported oil.

Chapter IV

RECOMMENDED GROUP STRATEGY

A. INTRODUCTION

As pointed out earlier, the numerical analysis of the future potential of new energy technologies is important, but it is only one aspect of strategic planning. Judgements must be applied to these analytical results which recognize that many aspects of our energy future are today uncertain. Strategic planning must attempt to take into account the influence of these inherent uncertainties. Of particular importance for strategy development are: (1) the uncertainty in predicting the results of RD & D, (2) the inability to predict with precision the future course of real energy prices; and (3) the impossibility of quantifying such matters as the social value and acceptability of certain technologies as they may be affected by societal preferences, trends in environmental requirements, government policies, etc. In the development of the strategy, an attempt was made to treat the implications of these unknowns, though in the case of the third factor, radical changes in society's preferences are not assumed.

As pointed out in Chapter I, the development of an energy RD & D strategy must be seen in terms of overall energy policy. To provide insights for RD & D policy making, two different policy assumptions were made that could affect technological needs, and that could lead to different sets of RD & D programme decisions. The two alternative strategies chosen for examination were:

i) Minimum Energy System Cost, and

ii) High Security of Energy Supply (interpreted as reducing oil imports for the IEA as a whole without incurring prohibitive additional costs).

The relative importance of individual technologies can be expected to vary among these two strategies. These alternative strategies are essentially represented by the PS-1 and the SP-4/PREM-2 scenarios described in Chapters II and III. The cost differences, import differences, and possible trade-offs were discussed in Chapter III.

B. CRITERIA FOR ESTABLISHING PRIORITIES

On the basis of the above considerations, the following criteria were considered of prime importance for strategy development:

- the potential amount and timing of each technology's contribution to energy supply or energy conservation;
- the degree to which the current state of knowledge makes it possible to estimate that the technology can successfully achieve the economic and technical goals assumed for it;
- the importance that the technology may have as a 'hedge' against uncertainties.

In order to apply these criteria, CRD developed the four categories of technology priorities set out in the following figure:*

In addition to these criteria, several other considerations were determined to be of importance in judging the priority to be assigned to individual technologies:

- **Early Contribution.** Certain technologies are projected to make an early contribution (in the

* There are other methods of ranking the importance of technologies than that used here. One of the participating countries (Germany) tested an alternative. This method involved grouping technologies into categories such as nuclear, coal liquefaction, conservation, etc. Following this, the analytical model was run omitting each technology group in turn. Where the omission of a technology resulted in surpassing a total fuel import limit, this technology was classed as Vital. Thus, depending on the effect of their omission on total oil imports, technologies were classed as Vital, Essential but Non-Vital, and less Essential. The results for Germany yield rankings very similar to those obtained from the CRD approach.

Figure 12
Group Strategy Priority Criteria

	Energy Contribution (Reference Case)*	Contribution as Hedge Against Future Uncertainty	Ability to estimate Performance Based on State of Knowledge
Priority One	High**		High
Priority Two	High		Less***
	or		
	Substantial**		High
Priority Three	Substantial		Less
	or		
		High or substantial	High
Priority Four		High or substantial	Less
	or		
	Low but Important Potential		

* The Reference Case for the strategy of Minimum Energy System Cost is PS-1-Oil Price B. The Reference Case for the strategy of High Security of Energy Supply is SP-4/PREM-2 which, as described in Chapter II, is based upon a premium cost above Oil Price B.

** Average annual energy contributions of 50 Mtoe or more were judged to be 'high'. Average annual energy contributions from 10 to 49 Mtoe were judged to be 'substantial'.

*** The technologies with which a lower level of knowledge is associated are identified 'K' on Figs. 13 and 14.

1980's). (Such technologies are marked with an 'E' in the priority Figures 13 and 14.)

— **Production of Liquid Fuels.** Some technologies provide a liquid fuel that permits direct substitution for oil. (Such technologies are marked with an 'L' in Figures 13 and 14.) While there are existing technologies - coal and nuclear - which can indirectly substitute for oil in electricity generation and certain heat requirements, technologies which permit direct substitution for oil in other applications are not available. Because of the importance of developing means of substitution in such uses, added emphasis is warranted.

— **Early Contribution and Production of Liquid Fuels.** Technologies which offered both early availability and provided a fuel directly substitutable for oil were assigned one higher priority position.

— **Long-Lead Development Time.** Since the analysis of potential utilization of new technologies covered a finite time period (1980-2020), the numerical results do not always adequately identify technologies which have major potential but whose development lead times are so long that their effect is felt late in the period. (Such technologies are marked 'LL' in the Figures 13 and 14 and, where warranted, are assigned a higher priority.)

— **Unique National Importance.** The IEA group strategy, serving among other things as a basis for developing and assessing national RD & D policies and plans, needs to recognize the possibility that some technologies may be of unique importance for oil import reduction in individual countries, even though their energy impact for the IEA as a whole may be limited. However, no such cases among the *new* technologies were identified, though as pointed out before, the application of existing technologies to countries which have not previously used them can require extensive development and demonstration.

It is also important to take into account expected environmental impacts in determining which technological opportunities to pursue. However, it was not possible to compare in quantitative or even semi-quantitative terms the different forms of environmental impact which arise from the various energy technologies. Nevertheless, it must be recognized that all forms of energy production have environmental impacts. These impacts must be taken into account during the development process and judged as to their acceptability before any new technologies are applied. This subject is treated more explicitly in Chapter V. In those cases in which known environmental or public safety issues call for further research and development, specific inclusion is made in the priority and indicative action recommendations.

It has also not been possible to analyse in detail the candidate technologies from the standpoint of their energy contributions relative to their costs of development. The CRD examined the priorities, which resulted from the above criteria, to judge whether the varying costs of development versus benefits should alter the priorities, and saw no reason to do so.

C. APPLYING THE CRITERIA

The criteria outlined above were applied for each of the two strategic directions outlined in Section A:

1. A Strategy of Minimum Energy System Cost

The results of the analyses show that, in a strategy aimed at minimizing the cost of the energy system, without accelerated development and introduction of new techologies, the technologies examined would make the energy contributions under varying assumptions as shown in Table 16.

When the priority categories and the other considerations described in Section B of this Chapter are applied, the priorities to be given to each given technology under this strategy are shown in Figure 13.

Table 16

IEA ENERGY RD & D STRATEGY PROJECT
AVERAGE ANNUAL UTILIZATION OF INDIVIDUAL TECHNOLOGIES (1990-2020) IN FOUR NON-ACCELERATED TECHNOLOGY MINIMUM SYSTEM COST SCENARIOS (1) (2)
(Mtoe)

Technologies Examined	Oil Prices B	Oil Prices C	Limited Fossil	Limited Nuclear	Oil Price B. Scenario % Change Between 2010 and 2020
End-Use					
Automotive transport systems	208	201	205	201	− 29
Alternative transportation fuels	74	90	21	73	+237
Conservation building (equipment)	264	268	272	244	+ 14
Electric auto	5	5	70	4	+ 56
Industrial conservation	254	254	256	249	+ 33
Residential & commercial solar heating	26	28	35	27	+ 32
Production					
Enhanced gas recovery	82	82	22	82	+ 33
Enhanced oil recovery	83	83	85	83	− 40
Geopressurized methane	20	19	0	37	− 40
Geothermal/Hydrothermal	31	31	32	31	+ 64
Hot dry rock	2	2	2	2	+ 70
Ocean power	5	5	14	15	+ 2
Shale and tar sands	143	148	143	143	+ 20
Solar electric	1	1	7	3	+5933 (3)
Wind power	5	9	24	22	− 3
Conversion					
Advanced converter reactors	46	46	67	21	+ 93
Breeder reactors	46	46	50	27	+158
Coal liquefaction	183	221	33	181	+109
Combined cycle	30	30	31	86	− 16
Fuel cells	1	1	6	2	− 72
Fuels from biomass	6	6	35	6	+ 39
Fusion	0	0	0	0	0
High caloric gasification	42	42	16	24	+ 46
Low-med. caloric gasification	7	7	1	6	+ 73
Magnetohydrodynamics	2	2	2	9	+1888 (3)
Non-fossil hydrogen systems	0	0	13	0	+1800 (3)
Underground gasification	1	1	1	1	+ 59

Note 1: All figures are rounded off to the nearest Mtoe.
Note 2: The oil price B and C scenarios include data from 15 participating countries. Some of the smaller countries did not compute the other two scenarios, and in these cases adjusted data from their Oil Price B scenario was used to provide a complete 15 country data set.
Note 3: 2010 contribution less than 0.5 Mtoe.

Figure 13

IEA ENERGY RD & D STRATEGY PROJECT TECHNOLOGY PRIORITIES IN A MINIMUM SYSTEM COST STRATEGY

Priority One	Priority Two	Priority Three	Priority Four
End Use			
Conservation - Building Equipment Automotive Transport Systems (E)	Residential and Commercial Solar Heating Industrial Conservation		Electric Auto - K
Production			
Enhanced Oil Recovery (L,E) Enhanced Gas Recovery (E) Shale and Tar Sands (L,E)	Geothermal (Hydrothermal)(E)	Geopressurized Methane - K Wind Power	Solar Electric - K Hot Dry Rock - K (LL) Ocean Power - K
Conversion			Low-medium Caloric Gasification
Coal Liquefaction - (L,E)	Combined Cycle (including low caloric gasification and pressurized fluidized bed)		Magnetohydrodynamics - K
Breeder Reactors (LL) Alternative Transport Fuels (L,E)	Fuels from Biomass (L,E)(3) High Caloric Gasification (E) Advanced Converter Reactors		Fuel cells - K Underground Coal Gasification - K Non-Fossil Hydrogen Systems - K Fusion - K (LL)
Key Supporting Technologies			
Environment protecting coal technologies (including atmospheric fluidized bed) Coal Mining Nuclear Reactor Safety Nuclear Fuel Cycle			

1. See Chapter IV, Section B, for definitions of priority categories.
2. Meaning of symbols used:
   K - Limited knowledge makes it less possible to predict that the technology will reach the goals attributed to it.
   L - Technology produces a liquid fuel that is directly substitutable for oil.
   LL - Long lead time limited the contribution the technology made over the period studied.
   E - Technology makes a contribution relatively early in the period studied.
3. This technology was moved up one priority category because it makes an early contribution (E) and it produces liquid fuels (L).

2. A Strategy of High Security of Energy Supply

The results of the analyses show that, in a strategy aimed at achieving high security of energy supply through accelerated development and introduction of new technologies and through a new technology cost premium, the technologies examined would make the energy contributions under varying assumptions as shown in Table 17.

When the priority categories and the other considerations described in Section B of this Chapter are applied, the priorities to be given to each technology under this strategy are shown in Figure 14.

Table 17

**IEA ENERGY RD & D STRATEGY PROJECT
AVERAGE ANNUAL UTILIZATION OF INDIVIDUAL TECHNOLOGIES (1990-2020)
IN FOUR ACCELERATED TECHNOLOGY SECURITY SCENARIOS AND IN A
SCENARIO THAT MAXIMIZES THE USE OF RENEWABLES (1)(2)**

(Mtoe)

Technologies Examined	Oil Prices B	Oil Prices C	Limited Fossil	Limited Nuclear	Maximize Renewables	Oil Price B Scenario % Change Between 2010 and 2020
End-use						
Automotive transportation systems	191	160	287	187	251	− 57
Alternative transportation fuels	168	171	15	166	90	+100
Conservation - building equipment	285	288	316	274	297	+ 12
Electric auto	12	12	51	12	9	+ 41
Industrial conservation	351	352	371	347	355	+ 18
Residential & commercial solar heating	56	57	78	57	78	+ 43
Production						
Enhanced gas recovery	92	92	28	123	70	+ 17
Enhanced oil recovery	89	89	89	89	90	− 76
Geopressurized methane	31	31	0	37	0	+ 64
Geothermal/Hydrothermal	39	36	41	42	43	+ 37
Hot dry rock	3	3	4	3	4	+ 77
Ocean power	3	4	13	15	21	+ 45
Shale and tar sands	192	192	192	192	184	+ 14
Solar electric	1	1	12	5	16	+1458 (3)
Wind power	4	6	24	23	55	+ 55
Conversion						
Advanced converter reactors	58	57	80	29	45	+ 88
Breeder reactors	68	68	76	32	51	+105
Coal liquefaction	298	301	19	278	198	+ 51
Combined cycle	32	35	22	92	7	− 14
Fuel cells	0	0	3	0	1	− 60
Fuels from biomass	9	11	60	12	14	+ 99
Fusion	0	0	0	0	0	0
High caloric gasification	41	41	18	56	52	+ 52
Low-med. caloric gasification	7	7	1	7	7	+ 73
Magnetohydrodynamics	5	5	2	11	2	+393
Non-fossil hydrogen systems	2	3	17	2	1	+691
Underground gasification	1	1	1	1	1	+ 69

Note 1: All figures are rounded off to the nearest Mtoe.

Note 2: The oil price B and Maximize Renewables scenarios include data from 15 participating countries. Some of the smaller countries did not compute the other three scenarios and in these cases adjusted data from their Oil Price B scenarios were used to provide a complete 15 country data set.

Note 3: 2010 contribution is less than 0.5 Mtoe.

Figure 14

IEA ENERGY RD & D STRATEGY PROJECT TECHNOLOGY PRIORITIES IN A HIGH SECURITY OF ENERGY SUPPLY STRATEGY

Priority One	Priority Two	Priority Three	Priority Four
End Use			
Conservation - Building equipment Automotive Transport Systems (E) Industrial Conservation Residential and Commercial Solar Heating		Electric Auto - K	
Production			
Enhanced Oil Recovery (L,E) Enhanced Gas Recovery (E) Shale Oil and Tar Sands (L,E)	Geothermal (Hydrothermal)(E)	Geopressurized Methane - K Wind Power	Ocean Power - K Solar Electric - K Hot Dry Rock - K (LL)
Conversion			
Advanced Converter Reactors Coal liquefaction - C (L,E) Breeder Reactors (LL) Alternative Transport Fuels (L,E)	High Caloric Gasification(E) Combined Cycle (including low caloric gasification and pressurized fluidized bed) Fuels from Biomass (L,E)(3)		Low-Medium Caloric Gasification Magnetohydrodynamics - K Fuel Cells - K Underground Coal Gasification - K Non-Fossil Hydrogen Systems - K Fusion - K (LL)
Key Supporting Technologies			
Environment protecting coal technologies (including atmospheric fluidized bed) Coal Mining Nuclear Reactor Safety Nuclear Fuel Cycle			

1. See Chapter IV, Section B, for definitions of priority categories.
2. Meaning of symbols used:
   K - Limited knowledge makes it less possible to predict that the technology will reach the goals attributed to it.
   L - Technology produces a liquid fuel that is directly substitutable for oil.
   LL - Long lead time limited the contribution the technology made over the period studied.
   E - Technology makes a contribution relatively early in the period studied.
3. This technology was moved up one priority category because it makes an early contribution (E) and it produces liquid fuels (L).

D. RECOMMENDED GROUP STRATEGY

The oil import implications of the two strategic directions described above are significantly different, while inspection of Figures 16 and 17 shows that there are very few differences between the two with respect to technology priorities. In view of the possibility of substantially reducing oil imports through accelerated development and deployment of promising technologies at relatively modest additional energy system costs, *it is the view of CRD that a strategy of High Security of Energy Supply be adopted.*

The Group Strategy has the following elements:

i) The pace of technology development considered essential if IEA energy policy objectives are to be realized;

ii) The priorities which individual generic technologies should have;

iii) Indicated actions proposing the next necessary and possible steps with respect to RD & D and commercialization of generic technologies; and

iv) Environmental considerations.

1. Accelerated Pace of Technology Development

Acceleration is likely to be a less expensive course of action than the consequences of a slower pace of development. At the same time CRD wishes to emphasize that the additional RD & D and commercialization costs of achieving such acceleration could not be determined in the course of this study.

The CRD recommends that the pace of new technology development and introduction be accelerated, with the degree of acceleration of individual technologies condidered on a case-by-case basis.

2. Generic Technology Priorities

Based upon the criteria and other considerations set out above, the CRD recommends that new energy technologies be accorded the following priorities:

Figure 15

IEA ENERGY RD & D GROUP STRATEGY
GENERIC TECHNOLOGY PRIORITIES

PRIORITY ONE

End-Use

Conservation - Automotive Transport Systems
Conservation - Building Equipment (mainly heat pumps)
Industrial Conservation Technologies
Residential and Commercial Solar Heating and Cooling

Production

Enhanced Gas Recovery
Enhanced Oil Recovery
Tar Sands and Oil Shale

Conversion

Advanced Converter Nuclear Reactors
Alternative Transport Fuels
Breeder Reactors
Coal Liquefaction

Key Supporting Technologies to make full use of existing technologies

Environment- Protecting Coal Technologies (including Atmospheric Fluidized Bed)
Coal Mining Technologies
Nuclear Reactor Safety Technologies
Nuclear Fuel Cycle Technologies

PRIORITY TWO

Production

Geothermal (Hydrothermal)

Conversion

Combined Cycle (including Low Caloric Gasification and Pressurized Fluidized Bed)
Fuels from biomass
High Caloric Gasification

PRIORITY THREE

End-Use

Electric Auto

Production

Geopressurized Methane
Wind Power

PRIORITY FOUR

Production

Hot Dry Rock - Geothermal
Ocean Power (Tidal, Wave, OTEC)
Solar Electric

Conversion

Fuel Cells
Fusion*
Low-Medium Caloric Value Gasification
Magnetohydrodynamics
Non-Fossil Hydrogen Production Systems
Underground Coal Gasification

---

* Fusion is ranked in Priority Four due to its late availability (2020). However, taking into account its potentially large contribution and long lead time, sufficient efforts should be made to develop this technology.

Application of Priority Rankings

It is emphasized that the above priority ranking is based upon the needs of IEA as a whole. As such, it is intended to portray the desired relative emphasis to be placed on the given technologies by the sum total of the energy RD & D programmes of IEA countries. It, therefore, will not correspond to the priorities of any one country. How the group strategy is expected to be used by the IEA is described in Chapter VII.

In viewing the relative priorities, the uncertainties in the data that have been used must be kept in mind. Thus, even the lower priority technologies require proper attention.

Furthermore, this strategy project has only addressed the development of new technologies; thus, the priority listing does not treat the effort which might be needed to apply available and elsewhere employed technologies in a country which has not yet used such systems. The effort required could be substantial, involving the development of a technological capability as well as physical infrastructure, and this could call for a high priority programme in any one country.

It is important to recognize that priority ranking should not be interpreted as translating directly into relative amounts of RD & D money to be spent among the individual technologies. Rather, it is intended to provide guidance in determining which technologies should have priority call for funding, considering their state of development. Thus, a high priority technology which is at an early exploratory phase may appropriately have a lower budget than a lower priority technology which is at the stage of demonstration. However, if the higher priority technology has need for additional funds to carry out a significant next step, it should have prior call on such funding.

Finally, it is noted that not all activities within a particular technology RD & D programme are of equal priority. Judgements as to importance are among the most vital in the planning and conduct of any RD & D programme.

3. Indicative Actions with respect to the next possible or necessary steps in the research, development, demonstration and commercialization of new technologies

Different technologies are typically at different stages of development. In this section the CRD indicates the next reasonable or appropriate steps in the process sequence from exploratory research to commercialization for each generic technology, taking into consideration the status of development already achieved. It is emphasized that these indicative actions are not necessarily intended to mean that direct governmental funding should be provided to support the work identified. Whether the support should be provided by government, the private sector, or jointly, and, if governmental action is warranted, whether it should be in the form of direct budgetary support or in the form of incentives or regulation will vary depending upon circumstances particular to individual country systems and the nature of individual technologies.

The indicative actions, in keeping with the purpose of this strategy, apply only to new technologies within these generic technology categories. In most cases there are several specific processes within each generic technology and some are already at a commercial stage; these cases are not covered below. Also they relate to the process(es) at the most advanced stage of technology development, and do not necessarily mean that each country will find its programme at the stage indicated or will find it appropriate to engage in any or all of the stages; some may not be working in the area at all or may contribute only a portion of the total effort needed to explore, develop and use the particular technology. Finally, it is emphasized that Figure 16 only indicates the current or next step warranted; in each case work in the preceding stages may well need to continue.

Figure 16

IEA ENERGY RD & D GROUP STRATEGY
INDICATIVE ACTIONS FOR GENERIC TECHNOLOGIES

i) **Proceed with the steps necessary to bring the following technologies to commercial use as early as practicable:**

End-Use

Conservation - Building Equipment (mainly heat pumps)
Industrial Conservation Technologies

Production

Enhanced Gas Recovery
Enhanced Oil Recovery
Tar Sands and Oil Shale

Key Supporting Technologies

Environment Protecting Coal Technologies (including Atmospheric Fluidized Bed)
Coal Mining Technologies
Nuclear Reactor Safety Technologies
Nuclear Fuel Cycle Technologies

ii) **Proceed to design, construction and testing of large-scale demonstrations in the following technology areas:**

End-Use

Alternative Transport Fuels
Conservation - Automotive Transport Systems
Residential and Commercial Solar Heating and Cooling Systems

Production

Geothermal/Hydrothermal

Conversion

Advanced Converter Reactors
Breeder Reactors
Combined Cycle, including Pressurized Fluidized Bed
Coal Liquefaction
Fuels from Biomass
High Caloric Gasification

iii) **Proceed with design, construction or manufacture, and testing of pilot scale experiments in the following technology areas:**

End-Use

Electric Auto

Production

Ocean Power
Solar Electric
Wind Power (large-scale)

Conversion

Fuel Cells
Low-medium Caloric Gasification
Underground Coal Gasification

iv) **Proceed with exploratory R & D and continuing assessment of results to determine whether and when to proceed to pilot plants or demonstrations:**

Production

Geopressurized Methane
Hot Dry Rock

Conversion

Fusion
Magnetohydrodynamics
Non-fossil Hydrogen Systems

Chapter V

ENVIRONMENTAL CONSIDERATIONS

It has been noted that all energy technologies impact the environment in some way. In providing for the energy needs of society, it is necessary that these impacts be kept to an acceptable level and that the environment be adequately protected. The determination of adequacy, however, is not simple. It is in part a technical question, on which there are varying degrees of knowledge about the various technologies. In part, this determination involves questions of value judgements. Furthermore it is necessary to consider the additional cost of an added level of environmental protection so that financial and manpower resources can be allocated most effectively among alternative needs.

Figure 17 is a simplified listing of the principal environmental factors which can be anticipated from the generic technologies studied. In a programme of energy research, development and demonstration, it is essential that such environmental factors be taken into account. In this project an attempt was made to do so at the strategic level by using environmental indicators as one of the optimizing factors. Ideally, such an analysis would reduce environmental effects to some limited number of common indices which could be used to compare directly the environmental effects of the technologies being studied.

Unfortunately, the environmental sciences have not progressed to the point where the diverse impacts of different technologies can be reduced to common, simple indicators. The question of environmental impacts of energy technologies is not a new one. A number of studies have been conducted which attempt to define and compare them. In one way or another all have fallen short of the achievement of that objective. Furthermore the degree of knowledge as to environmental effects varies with the technologies. In general, more is known about the effects of existing types of energy systems than is known about those technologies which are at very early stages of development. Indeed, it is often necessary to conduct RD & D, and even build systems, in order fully to understand the environmental implications. Thus, it was not possible to follow this idealized approach.

Yet, it remains essential that environmental needs be fully accounted for in the conduct of energy RD & D. In order to provide adequate protection of the environment, it is necessary that:

— Each energy RD & D activity take into account the environmental implications of the technology and the means of reducing them to an acceptable level. Where the environmental effects are not known or are poorly understood, the research activities should specifically include elements to gain adequate knowledge of these factors.

— Before a new technology is extensively deployed, a specific assessment as to the environmental effects of deployment of the technology should be made.

While it was not possible to include an environmental optimization in the strategy project, environmental factors were included in two specific ways:

1. For certain of the existing and new technologies, the technology characterizations include allowances for the added costs and altered performance characteristics which are necessary to meet the various existing and projected national standards for environmental protection. The characterizations have included such items as abatement of effluents, waste management, and land restoration. It is noted that knowledge of these factors is less exact for the more speculative technologies which are at an early stage of development.

2. Where there are known problems of an environmental or public safety nature, recommendations on related R & D have been made in the priority and programme statements. Here again the problems are better known for those technologies with which there has been greater experience.

Figure 17

IEA ENERGY RD & D STRATEGY PROJECT : PRINCIPAL ENVIRONMENTAL FACTORS

Note: This chart identifies the principal environmental factors associated with the generic technologies. The symbols "L" (Large); "M" (Moderate); "S" (Small); and "O" (None), are intended to provide a semi-quantitative measure of the relative size of the particular environmental factor among the several technologies. However, the effects of some of these factors are subject to minimization by proper design, technology and management. Others are characteristics, the effects of which cannot be altered by such means. For the latter, the column heading is identified with the term "inherent". In the case of supply technologies, assessments apply to the complete fuel cycle systems; however, environmental effects of the manufacture of materials and components, which are used in such systems are not included and in some cases can be significant.

	AIR EMISSIONS				CONSUMPTION		WASTE MATERIALS			
	CO_2 (Inherent)	SOx	NOx	Particulates	Water Use	Land Use (Inherent)	Quantity	Toxicity	Radio-activity	Notes
End-Use										
Automotive transportation systems	L	S	L	M	M	?	O	O	O	System Dependent
Alternative transportation fuels	L	S	L	M	M	S	O	O	O	
Conservation building equipment	O	O	O	O	O	O	O	O	O	Energy supply to operate equipment
Electric auto	Depending upon method for charging system									
Industrial conservation	(Process specific)									
Residential and commercial solar heating	O	O	O	O	S	S	S	S	O	
Production										
Enhanced gas recovery	L	S	S	O	?	M	S	S	O	
Enhanced oil recovery	L	S	S	S	L	M	S	S	O	
Geopressurized methane	L	L	S	S	L	M	S	S	O	
Geothermal/Hydrothermal	O	L	O	O	S	L	M	M	O	
Hot dry rock	O	S	O	O	L	L	M	M	O	
Ocean power	O	O	O	O	S	S	S	S	O	Disruption of marine ecology
Shale and tar sands	L	L	S	L	L	L	L	M	S	
Solar electric	O	O	O	O	L-M	L	M	S	O	
Wind power	O	O	O	O	S	L	S	S	O	
Conversion										
Advanced converter reactors	O	O	O	O	S	M	S	L	L	
Breeder reactors	O	O	O	O	S	M	S	L	L	
Coal liquefaction	L	L	M	L	L	L	L	M	S	
Combined cycle low quality gasification	L	L	L	L	S	L	L	S	S	
Fuel cell (coal gas)	L	S	S	S	L	L	S	M	S	
Fuels from biomass	O	O	M	S	L	L	M	S	O	
Fusion	O	O	O	O	S	M	S	L	L	
High caloric gasification	L	L	M	L	L	L	L	M	S	
Low-med. caloric gasification	L	L	M	L	L	L	L	M	S	
Magnetohydrodynamics	L	M	M	S	S	L	L	M	S	
Non-fossil hydrogen systems	O	O	O	O	S-L	S	?	?	?	Dependent on method
Underground gasification	L	L	M	S	L	L	S	S	S	

Chapter VI

SOME NON-TECHNOLOGY POLICY ISSUES AFFECTING THE INTRODUCTION OF NEW ENERGY TECHNOLOGIES

Up to this point, the report has only discussed technical, economic, and environmental factors. These factors were judged to be the most important and influential in the selection of technology priorities and strategy thrust.

However, the pursuit of any strategy and the eventual deployment of new technology can be affected by a variety of non-technological and institutional issues. It has not been possible to analyse directly the quantitative influence of such factors on possible future energy systems development or on the introduction of new technologies into the system, though to some degree they were introduced in a judgemental manner in establishing the technology characterization inputs to the analysis and in the assessment of the results.

Furthermore, some are often secondary to the technical, economic, and environmental issues. Experience has usually shown that as a technology begins to look very favourable with regard to the three primary factors, it becomes easier to find solutions to the non-technological issues. The effort to minimize such problems generally begins to be applied when a technology is judged to be truly competitive and environmentally sound.

Yet the non-technological issues remain as hurdles to be overcome, requiring appropriate attention throughout the development stages. CRD believes that adequate attention to them is essential for the achievement of the potential of new technologies. The objective must be to remove those barriers which might prevent or slow down a technology from achieving that position. A sampling here is intended to identify some of the more crucial concerns.

Rational Pricing Signals

Many countries have chosen to intervene in energy markets by establishing pricing policies that satisfy other national objectives. These actions have the effect of distorting the relative costs of different forms of energy as seen by the consumer, and can consequently affect the entry of new energy resources and systems into those markets. Often, these pricing policies have the effect of continuing dependence on existing energy sources and hindering the introduction of new sources of energy.

Of the non-technological issues, the establishment of rational (or correct) prices, i.e., prices that are not below world market levels, is probably the most important in terms of any strategy. The analytical work supporting this strategy, giving indications of the possible deployment of a large array of technologies, is based on the relative economics being correctly expressed.

If pricing policies keep costs of any or all energy forms to consumers below replacement costs of energy for any length of time, the future implementation of RD & D products will be delayed. In particular, it is imperative that governments not maintain incorrect pricing of oil for any extended time.

Regulatory Requirements

Governments, at both the national and local level, often impose regulations on the conduct of business within their countries. Because these regulations are intended to protect both businesses and individual consumers in areas such as health, safety, anti-trust, geo-political interests and economic security, the potential is present for conflicting and confusing regulations which affect new energy development.

Three primary actions are required to alleviate potential impediments: technologies must be developed to ensure compliance with regulations; governments must ensure stability of regulations over reasonable periods of time; and governments must demonstrate a willingness to prevent and to remove conflicting regulations. On the other hand, regulations may in some cases be necessary to facilitate introduction of new technologies when market forces or other forms of economic signals do not adequately portray the national objectives.

Foreign Trade Policy

Countries use a variety of instruments, such as tariffs, quotas, development funds, and patent policies, to foster economic trade advantages. Unless early, careful attention is given to these instruments, successful deployment of some new energy technologies may be made difficult in individual IEA countries.

Energy export and import policies among IEA countries will become increasingly important. The strategy analysis has shown a trend towards significant increases in the use of coal, uranium, and liquid substitutes. Only a few IEA countries are blessed with large quantities of these resources. Yet, the successful achievement of cost and security goals is dependent upon greater availability in the multi-national sense.

Countries will face trade-offs between a stable international energy economy on the one hand and the desire to reap the benefits of resources and high technology developed within their countries on the other. These issues manifest themselves in restrictive import or export measures, patent policies, and licensing arrangements. Those priority technologies whose development is affordable only to the major countries, such as coal liquefaction and gasification, advanced reactors, and new automobile engines, are key to the strategy. They must also be made available to technology importing countries in the least restrictive manner possible. Co-operation through the IEA to minimize such barriers is essential.

Socio-Economic Impacts

Some of the new energy technologies are sufficiently compatible with the existing energy system that they will not require much change in the surrounding socio-economic structure. A few technologies, which happen to show some of the largest potential contributions, will require significant adjustments. Major potential impacts are economic dislocation through structural changes in the composition of the energy system, municipal services to support the growth of industry in particular regions, and cultural impacts on the indigenous population.

Problems of this nature have been treated by governments in the past. These issues, while not necessarily absolute barriers to new developments, can be delaying obstacles if proper advance planning has not been undertaken. The solutions usually take the form of additional flows of money and services into the affected area. Coordination with the non-energy parts of governments and with local officials is required.

Supporting Resource Availability

The development and growth of new energy forms will require that many supporting resources be available. These resource issues include availability of land (energy leasing policies), water rights, an infrastructure for development and distribution of the products, ancillary raw materials, and skilled labour.

The assurance of these resources requires collaborative action between governments and industry. Land and water rights for energy development are in competition with agricultural development and industrial development, with allocations sometimes established prior to having taken into account particular needs of new energy technologies. Development of an infrastructure, raw materials, and skilled labour has typically been adequate if appropriate signals are given to the industrial sector. However, delays in availability have and will be experienced if there is insufficient attention to this problem.

With regard to this strategy, particular emphasis must be given to the supporting resource requirements for coal development. Because coal plays such an important role in most countries, either through direct use or transformation, inducements will be necessary to attract a qualified labour force and to construct an adequate transportation and ports system.

Chapter VII

USE OF THE RECOMMENDED GROUP STRATEGY AND FUTURE ACTIVITIES

A. USE OF THE GROUP STRATEGY

As pointed out in Chapter I, one of the purposes of the IEA strategy is to provide a basis for:

- developing and assessing national RD & D policies and plans, and
- achieving the most effective relationship between the RD & D activities of the Member countries.

At the same time it has been emphasized that the IEA strategy is not intended to be, nor can it be, identical with individual national energy RD & D strategies and programmes. Therefore, in carrying out the above purposes in a manner consistent with the inherent distinction between an international strategy and a national one, the CRD expects to use the IEA strategy as one of the factors in its periodic reviews of national RD & D programmes.

1. At the National Level

a. Member countries are urged to accelerate their efforts to develop and bring to commercial use important new energy technologies.

b. As national energy RD & D strategies are developed and evolve, the Group Strategy priorities should be one of the criteria used in setting national RD & D priorities. The indicative actions are intended to assist countries in planning individual technology programmes. In its periodic reviews of national RD & D programmes, CRD will consider the insights provided by the IEA group strategy in evaluating the strategies of individual countries, taking into account relevant national circumstances such as the following:

- national energy policies
- existing technological capability
- natural resources
- industrial development objectives
- overall economic resources.

2. At the International Level

a. Using information collected in the course of its annual review of Member countries' energy RD & D programmes, the CRD intends to compare the aggregate of energy RD & D efforts of IEA countries as a group with the group strategy, and to make recommendations accordingly.

b. In its collaborative project activities, the CRD will consider the importance of the technologies as portrayed in the strategy, recognizing that there are additional factors which influence the potential for, and value of, such collaboration.

B. FUTURE ACTIVITIES

Prior to, or in the course of this project, the fifteen countries participating in the systems analysis effort initiated the development of their national strategies. In many cases this work is still underway. CRD hopes that those countries, as well as the other IEA countries which did not participate, will carry such programmes to completion.

In addition, as noted earlier, there are many questions that the work done so far does not address, e.g. further studies of the sensitivity of projections to different assumptions and more detailed information on processes within generic technology categories, particularly in end-use efficiency. Continuation of analytical work in a co-ordinated manner is planned by some of the participating countries.

Finally, since an IEA RD & D strategy, as well as national energy RD & D strategies, must be recognized as evolutionary in character, adapting to new information and insights that arise externally as well as internally to the programmes, CRD intends to re-examine the strategy periodically in due course.

Project Bibliography

1. *Final Report of the IEA Steering Group for Energy R & D Systems Analysis*, to be published in 1981.

2. *User's Guide to MARKAL (KFA Version)*, May 1980.

3. *Documentation of the MARKAL Model (KFA Version)*, May 1980.

4. *Energy Technology Data Handbook* (2 vols) (KFA Report No. Juel-Spez-70), April 1980.

5. *IEA Energy Systems Analysis Project: Technology Review Report*, BNL Report 27074, Dec. 1979.

6. Abilock, H. and Fishbone, L.G., *User's Guide for MARKAL (BNL version)*, BNL Report 27075, Dec. 1979.

7. *An Initial Multi-National Study of Future Energy Systems and Impacts of Some Evolving Technologies*, BNL Report 50641 and KFA Report Juel-1406, March 1977.

8. *Towards an International Strategy for Energy Research and Development*, IEA, Sept. 1977.

ANNEX I

IEA PRINCIPLES FOR ENERGY POLICY ADOPTED BY THE GOVERNING BOARD AND EXCERPT FROM IEA MINISTERIAL LEVEL GOVERNING BOARD MEETING COMMUNIQUE OF 22nd MAY, 1980

Principles for Energy Policy

1. Further development by each Participating Country of national energy programmes and/or policies which include the objective, formulated as specifically as possible, of reducing in absolute terms or limited future oil imports through conservation of energy, expansion of indigenous energy sources and oil substitution.

2. Constant and careful attention to important environmental, safety, regional and security concerns to which the production, transportation and use of energy give rise, and improvement of the speed and consistency of public procedures for resolving conflicts which may exist between these concerns and energy requirements.

3. Allowing domestic energy prices to reach a level which encourages energy conservation and development of alternative sources of energy.

4. Strong reinforcement of energy conservation, on a high priority basis with increased resources, for the purpose of limiting growth in energy demand relative to economic growth, eliminating inefficient energy use, especially of rapidly depleting fuels, and encouraging substitution for fuels in shortest supply, by implementing vigorous conservation measures in various sectors along lines which include the following elements:

- pricing policies (including fiscal measures) which give incentives to conservation;
- minimum energy efficiency standards;
- encouragement and increase of investment in energy saving equipment and techniques.

5. Progressive replacement of oil in electricity generation, district heating, industries and other sectors by:

- discouraging the construction of new exclusively oil-fired power stations;
- encouraging the conversion of existing oil-fired capacity to more plentiful fuels in electricity, industrial and other sectors;
- encouraging the necessary structural adjustments in the refinery sector in order to avoid an excess of heavy fuel oil;
- directing efforts to the reduction of the use of heavy fuel oil as a primary energy source in those sectors where efficiency is low.

6. Application of a strong steam coal utilization strategy and active promotion of an expanded and reliable international trade in steam coal, composed of the following elements:

- rapid phasing-in of steam coal as a major fuel for electrical power generation and in industrial sectors;
- further development of steam coal policies within producing, exporting and consuming IEA countries to support increased utilization by enhancing market stability through reliable and increased export and import flows under reasonable commercial terms;
- development of policies to remedy anticipated infrastructure bottlenecks.

7. Concentration of the use of natural gas on premium users' requirements, and development of the instructure necessary to expand the availability of natural gas.

8. Steady expansion of nuclear generating capacity as a main and indispensable element in attaining the group objectives, consistent with safety, environmental and security standards satisfactory to the countries concerned and with the need to prevent the proliferation of nuclear weapons. In order to provide for this expansion, it will be necessary through co-operation to assure reliable availability of:

— adequate supplies of nuclear (uranium and enrichment capacity) at equitable prices;

— adequate facilities and techniques for development of nuclear electricity generation, for dealing with spent fuel, for waste management, and for overall handling of the back-end of the nuclear fuel cycle.*

9. Stronger emphasis on energy research, development and demonstration, including collaborative programmes, more intensive national efforts and greater coordination of national efforts, in order to make energy use more efficient and to meet future energy requirements. Each Participating Country should contribute to energy technology development, with emphasis on: (a) technologies which can have relatively near-term impact, (b) policies which facilitate the transition of new energy technologies from the research and development phase to the point of utilization, (c) technologies for broadly applicable renewable energy sources, and (d) investigation of whether there are technological possibilites for significant contributions from other renewable resources, through :

— providing the fullest possible financial support for energy research, development and demonstration;

— increasing participation in international collaborative projects to extend the effectiveness of funds available;

— encouraging investment in energy/technology development by appropriate incentives;

— ensuring that R & D policies remain consistent with, and supportive of, the objective of ongoing energy policy.

10. Establishment of a favourable investment climate which encourages the flow of public and private capital to develop energy resources by appropriate pricing policies, by minimising uncertainties about the general directions of energy and other policies such as mentioned in Principle 2, and by providing government incentives where necessary, in order to:

— give priority to exploration activities including those in offshore and frontier areas;

— encourage rates of exploration and development of available capacities which are consistent with the optimum economic development of resources.

11. Providing in energy policy planning for alternative means, other than increased oil consumption, for meeting any development of supply shortfall or failure to attain conservation objectives, taking into account the appropriate requirements of economic development and social progress.

12. Appropriate co-operation in the field of energy, including evaluation of the world energy situation, energy research and development and technical and financial requirements, with developed or developing countries or international organizations.

From Ministerial Meeting Communique of 22nd May 1980

Ministers noted that an IEA RD & D Group Strategy has been developed. They concluded that the Governing Board at official level will pursue the strategy's accelerated scenario, which minimizes oil imports for the IEA as a whole. They agreed that IEA countries will use the IEA RD & D Group Strategy as a guide for setting national priorities and funding levels as well as for IEA collaborative project priorities. The Committee on Energy Research and Development will closely monitor and periodically consider the extent to which aggregate national RD & D efforts are consistent with the Group Strategy. The Coal Industry Advisory Board is invited to provide recommendations as to which new technologies should be pursued in order to further speed up expanded production and use of coal. (para. 16)

* The following Delegations expressed individual positions regarding Principle No. 8, as set forth in the Conclusions of the Meeting of the Governing Board which adopted this Decision: Denmark, The Netherlands, New Zealand, Norway, Spain, Sweden.

ANNEX II

TECHNOLOGY CHARACTERIZATION SUMMARY

The attached table lists reference values for some significant parameters which characterize new technologies that have been used in the IEA Energy R & D Systems Analysis Project. The actual data used in the calculations for individual countries might differ from the reference data as a consequence of special pollution abatement requirements, local economic and geographic factors, and other country-specific factors which impact unit sizes, costs, and efficiencies. Detailed information for each technology for each participating country can be found in the Technology Review Report.*

The table lists "conversion" technologies (CON), which include electricity generation and the production of heat for industrial processes and district heating grids, and "process" technologies (PRC), which include transformations from one fuel type to another (oil refineries, coal gasification, etc.).

General Ground Rules:

The project has adopted several basic ground rules:

- All costs* were expressed in constant 1975 US dollars. Costs originally in other national currencies have been deflated to 1975 according to national experiences and extrapolations and were converted to dollars at 1975 average exchange rates. In a few cases, adjustments were required to account for currency revaluations.
- The capital investment costs include interest during construction calculated at 6 per cent per annum in real terms.
- Taxes, subsidies and profits have not been included in the capital and operating costs calculations, these being viewed as internal exchanges within each national system. This rule does not apply to commodities offered for export.
- Fixed operating and maintenance costs include insurance premiums and other specified real expenses (excluding taxes) required by law in the individual countries (e.g., escrow payments for waste disposal or decommissioning.)
- Capacities are expressed in units of megawatts net electrical output (MWe) for (CON) systems, and Mtoe per year for (PRC) systems.
- The energy contents of coal, gas and oil-based fuels are expressed in terms of *net* calorific value, i.e. the heat of condensation of combustion product H_2O is not included in the calorific content of the fuel.
- Efficiences are *net* efficiencies.
- The contribution to total primary energy of nuclear and renewable systems is calculated in terms of fossil-fuel equivalent (FEQ). This is the fossil-fuel that would have been consumed to supply the same services. Numerical values of FEQ for electrical systems are those listed in Table B2, page 103 of OECD, *World Energy Outlook*, 1977, for the years 1980 and 1985.

* Costs are not shown in this Annex because they change with time. They are contained in the *Technology Review Report*, December 1979, Brookhaven National Laboratory.

Values for later years are calculated by each country. Values of FEQ for non-electrical systems are calculated from project data. *The thermal efficiencies for nuclear and renewable systems are not used by MARKAL for any calculation* other than net heat release to the environment. In the case of nuclear systems, MARKAL requires input data which gives the tonnes of fabricated fuel required to produce one petajoule of net electric energy output.

Procedures Used in the Technology Review:

The data review process consisted of the following steps:

1. The project staff at BNL/KFA made a preliminary comparison of the input data of the countries at each laboratory (Japan, Sweden and the United States at BNL, Germany and United Kingdom at KFA).

   This served to identify major discrepancies, errors and systematic differences.

2. Each national participant consulted resources within his home country to obtain additional advice and opinion concerning specific technologies where there were reasons for doubt.

3. The MARKAL data elements which describe each technology were submitted wherever possible for critical review to experts within BNL and KFA and to outside experts. Many of the individuals and organizations who have contributed to this step of the review are cited in the acknowledgement section, page viii (Ref. 1). For some technologies we have not been able to obtain critical review from outside experts, in some cases because of proprietary information considerations or because the particular information desired is not obtainable as such. Data values necessarily rely on personal judgements either of an expert of the project staff when expert opinions differed.

4. A technology review workshop was held at KFA-Jülich from September 18-22 1978, to compare all available information and to select "reference input data". The workshop was attended by participants from each of the five Phase A countries and assisted by other members of the KFA IEA Staff. Each technology and various other data elements were assigned to individual participants for investigating unanswered questions and for documentation of sources. A *Technology Review Report* was the product of the workshop.

5. Additional critical comments were received following circulation of the *Technology Review Report* to Steering Group Delegates and participating governments. These resulted in a reiteration of steps 1-3 above, where appropriate.

6. In June 1979, data from nine countries were added to the Technology Review tables. This process raised a few additional questions which have subsequently been investigated by measures similar to steps 1-3.

It should be emphasised that the technology data review has been a continuing process up until the time of the final runs for each country.

'Reference Values' for Characterization of New Technologies:

At the time of the September 1978 technology review workshop at KFA, individuals and/or subcommittees were assigned the tasks of preparing sets of "reference values" for the input data elements which were expected to be more or less independent of the country in which the system was deployed. The draft reference values were reviewed within the project and in some cases referred back to outside experts for final comments.

A set of reference values was prepared for each new (PRC) or (CON) technology which has been included in the data base of two or more countries. Included in the data set are:

- Start year (the period of first commercial availability. The start year should be that of a country where the technology is expected to be available first. Countries not pursuing active R & D at the present time will use a later start year based on an assumed import of a successfully demonstrated technology by the lead country.)
- Technical Lifetime
- Availability Factor
- Efficiency (net efficiency for fossil conversion systems, or output fractions having more than one output. Note that this datum is not used for nuclear and renewable systems, although it is sometimes cited. Nuclear systems require an input datum which lists the tonnes of fabricated fuel required to produce one petajoule of net electric energy output.)
- Specific Investment Cost (capital cost including interest during construction)

- FIXOM (fixed annual operating and maintenance costs)
- VAROM (variable annual operating and maintenance costs)

The UNIT SIZE which was the basis for the system modelling is cited, but is not used in the MARKAL calculations.

The other data elements required to characterize a technology in the MARKAL input data are not included in the intercountry comparison because they are all country specific.

The tables which follow summarize several of the more important parameters. More detail can be found in the *Technology Review Report* cited above.

TABLE II-1 : **CONDENSED REFERENCE TECHNOLOGY DATA**

TECHNOLOGY	UNIT SIZE	REFERENCE START YEAR*	EFFICIENCY %	NOTES (see p. 73)
END-USE				
AUTOMOTIVE TRANSPORT SYSTEMS:				
Nuclear Powered Ship, PWR (Marine)	1 Mtoe/yr	2000	–	Jap
(See also Table II-2)				
ALTERNATIVE TRANSPORTATION FUELS:				
(See: Coal Liquefaction; Fuels from Biomass; Non-fossil Hydrogen Systems. Also see Table II-2)				
Production of Methanol from Natural Gas	3 Mtoe/yr	1985	60	NZ
Conversion of Methanol to Gasoline (MOBIL)	1 Mtoe/yr	1985	86	NZ
Production of Gasoline from Natural Gas (SYNTHOL)	2 Mtoe/yr	1985	43	NZ
CONSERVATION BUILDING - EQUIPMENT:				
Seasonal Low-Temperature Heat Storage, Underground (See also Table II-2)	8 Mtoe/yr	1990	70	Ita
ELECTRIC AUTO:				
(See Table II-2)				
INDUSTRIAL CONSERVATION:				
Binary Cycle Cogeneration, Medium Temperature Heat	1 MWe	1990	5	Ita, 7
Industrial Cogeneration, Gas Turbine	80 MWe	1985	30	US, 7
Industrial Cogeneration, Coal-Steam Back Pres.Turbine	30 MWe	1985	13	7
Industrial Cogeneration, Diesel	–	1985	35	7
Medium Temperature Heat from Solar	1 Mtoe/yr	1980	–	Ita
(See also Table II-2)				
RESIDENTIAL AND COMMERCIAL SOLAR:				
(See Table II-2)				
PRODUCTION				
ENHANCED GAS RECOVERY:				
(Treated as domestic extraction via supply curve)		1985		3
ENHANCED OIL RECOVERY:				
(Treated as domestic extraction via supply curve)		1980		2
GEOPRESSURIZED METHANE:				
(Treated as domestic extraction via supply curve)		1990		US
GEOTHERMAL/HYDROTHERMAL:				
Geothermal/Hydrothermal Power Plant (existing technology in several countries, historical data used for each country)				
HOT DRY ROCK:				
Dry Geothermal (Hot Rock) Power Plant	100 MWe	1990	–	
District Heating Plant, Geothermal		1980	–	Jap

TABLE II-1 : CONDENSED REFERENCE TECHNOLOGY DATA (CONTINUED)

TECHNOLOGY	UNIT SIZE	REFERENCE START YEAR*	EFFICIENCY %	NOTES (see p. 73)
OCEAN POWER:				
Wave Central Electric Power Plant	2 MWe	1990	–	
Ocean Thermal Gradient Electric Power Plant	100 MWe	2000	–	
Tidal Electric Plant	800 MWe	1985	–	
SHALE OIL AND TAR SANDS:				
Tar Sands Processing, Flexicoking	9 Mtoe/yr	1980	82	Can
SOLAR ELECTRIC:				
Decentralized Solar Photovoltaic	10 kWe	1985	–	
Central Solar Thermal Electric	150 MWe	2000		
Central Solar Photo-Voltaic Power Plant	1 MWe	1990		
WIND POWER:				
Wind Turbines, Central Electric Power Complex	2 MWe	1985	–	
Local, Wind Electric Generator	0.02 MWe	1985	–	
CONVERSION				
ADVANCED CONVERTER REACTORS:			(Note 6)	
ATR Nuclear Power Plant	1000 MWe	1980		Jap
CANDU Nuclear Power Plant, Enriched Fuel	1000 MWe	1990		Can
CANDU Nuclear Power Plant, Highly Enriched Fuel	1000 MWe	2000		Can
AGR Nuclear Power Plant	1000 MWe	1980		UK
HTR Nuclear Power Plant	1000 MWe	1995		
Nuclear Process Heat, VHTR, Enriched Uranium Fuel	2 Mtoe/yr	2000		Jap
Nuclear Process Heat, VHTR, U33 Fuel	2 Mtoe/yr	2010		Jap
Nuclear Process Heat, CANDU-ORC	2 Mtoe/yr	1990		Can
VHTR Nuclear-Coal, Reducing Gas		2000		Jap
(See Technology under High Caloric Gasification for nuclear systems used for coal gasification				
BREEDER REACTORS:				
LMFBR Nuclear Power Plant	1000 MWe	1995	(Note 6)	
COAL LIQUEFACTION:				
Hard Coal Liquefaction, Hydrogenation	9 Mtoe/yr	1995	65	1985 US
Hard Coal Liquefaction, Fischer-Tropsch	8 Mtoe/yr	1995	58	1,1985 US
Hard Coal Liquefaction, SRC-2 Process	2 Mtoe/yr	1995	70.3	Jap.
Methanol Production from Hard Coal	2 Mtoe/yr	1990	46	1,1985 US
Brown Coal Liquefaction, Hydrogenation	2 Mtoe/yr	1995	55	
Brown Coal, Methanol Production by LURGI		1995	46	1,1980 Can
COMBINED CYCLE, LOW QUALITY GASIFICATION: (including Fluidized Bed Combustion)				
Hard Coal Combined Cycle Power Plant	1200 MWe	1990	44.9	1
Pressurized Fluidized Bed Power Plant	1000 MWe	1995	43	
Atmospheric Fluidized Bed Power Plant	1000 MWe	1990	38	
Atmospheric Fluidized Bed Industrial Boiler for Process Steam	28 MWth	1985	70	US
FUEL CELL (COAL DERIVED FUEL):				
Gas Fuel Cell	26 MWe	1985	40	
FUELS FROM BIOMASS:				
Coal/Municipal Waste, Steam Electric	350 MWe	1980	32.2	4, US
Coal/Municipal Waste, Steam Electric	350 MWe	1980	32.2	US
Biomass Steam Electric Power Plant	50 MWe	1985	31	1980 Can, Nor, Swe
Methanol from Wood	1 Mtoe/yr	1985	40	
Gas from Wastes		1985		Ita

TABLE II-1 : CONDENSED REFERENCE TECHNOLOGY DATA (CONTINUED)

TECHNOLOGY	UNIT SIZE	REFERENCE START YEAR*	EFFICIENCY %	NOTES (see below)
FUSION:				
Nuclear Fusion Power Plant	750 MWe	2020	(Note 6)	2010 Can
HIGH CALORIC GASIFICATION OF COAL:				
Hard Coal, Nuclear Hydrogasification	5 Mtoe/yr	1995	43.1	
Hard Coal Gasification, Nuclear Combined Hydro/Steam	3 Mtoe/yr	1995	69.5	
Hard Coal, High Btu Gasification, LURGI-Slagging	5 Mtoe/yr	1990	60	
Brown Coal, Nuclear Hydrogasification	4 Mtoe/yr	1995	48.4	
Brown Coal, Nuclear Steam Gasification	4 Mtoe/yr	1995	56.8	
Brown Coal, High Btu Gasification, LURGI-Slagging	4 Mtoe/yr	1990	65	1,1985 Can
LOW TO MEDIUM CALORIC GASIFICATION OF COAL:				
Hard Coal, Medium Btu Gasification	1 Mtoe/yr	1990	60	1
Hydrogen Production from Hard Coal	2 Mtoe/yr	1980	66	Can
Brown Coal, Medium Gasification, Winkler Process	2 Mtoe/yr	1980	60	
Prod. of Reducing Gas from Coal & Nuclear Process Heat		2000	52.3	
MAGNETOHYDRODYNAMICS:				
Coal MHD Electric Power Plant	1932 MWe	2005	48	
NON-FOSSIL HYDROGEN SYSTEMS:				
Hydrogen Production by Electrolysis of H_2O	.002 Mtoe/yr	1985	76	1980 Can. Nor
Hydrogen Production by Thermochemical Splitting using Nuclear Process Heat	1 Mtoe/yr	2000	38	5
UNDERGROUND COAL GASIFICATION				
Underground Gasification with Combined Cycle Electric Power Plant	170 MWe	1990	22.7	
In Situ Coal Gasification	2 Mtoe/yr	1985	58	

* REFERENCE START YEAR is defined as the year in which the reference technology could begin commercial operation (date available plus lead time for design and construction). In some cases, for special reasons, the technology could start operation in one country at an earlier date. Such exceptions are listed in the Note column.

NOTES FOR TABLE II-1

Technologies used in only one country are noted by country abbreviations:

Aus	Austria	Jap	Japan
Bel	Belgium	NZ	New Zealand
Can	Canada	Nor	Norway
Den	Denmark	Spa	Spain
Ger	Germany	Swe	Sweden
Ire	Ireland	UK	United Kingdom
Ita	Italy	US	United States

1. Efficiency improves in later years.
2. "Enhanced Oil Recovery" is defined as tertiary oil recovery using steam, CO_2 miscible, micellar polymer and in-situ combustion processes. Secondary recovery using water flooding is included under conventional oil recovery. Enhanced Oil Recovery was divided into three production cost groups: $7.50/bbl, $12.50/bbl, and $20.00/bbl, each with a defined total resource.
3. "Enhanced Gas Recovery" is from Devonian Shales, Coal Seams, and Western Tight Sands. The resources have been divided into two production cost classifications: $2.00/Mcf and $6.00/Mcf.
4. Includes sulphur abatement equipment.
5. Unit size and costs refer to hydrogen *output*.
6. The MARKAL model does not make direct use of the thermal efficiency of the nuclear system. Instead the input datum lists the tonnes of fuel required to produce one petajoule of net electric energy putput. This value is embodied in the fuel cycle parameters which cannot be summarized in simple form.

7. The cogeneration systems produce both electricity and industrial process heat. The units are sited in industrial complexes. For details about heat production the reader is referred to the Technology Review Report.

8. Coupled production plants produce both electricity and low temperature heat for district heating. The reader is referred to the Technology Review Report for details on performance.

TABLE II-2: EXAMPLES* OF NEW END-USE TECHNOLOGIES USED IN THE NATIONAL DATA BASES OF THE IEA ENERGY SYSTEMS ANALYSIS PROJECT

TECHNOLOGY	START**	NOTES
END-USE		
AUTOMOTIVE TRANSPORT SYSTEMS:		5
Private Auto, Improved Otto-cycle	1980	3
Private Auto, Improved frame	1980	3
Private Auto, Improved Diesel	1980	3
Private Auto, Improved LPG	1980	3
Fuel Cell Powered Auto	2005	
Stirling Engine, Private Auto	1990	
(See Table II-1)		
ALTERNATIVE TRANSPORTATION FUELS:		
Private Auto, Methanol Fuels	1985	4
Private Auto, Hydrogen Fuel	1995	
Truck and Bus, Hydrogen Fuel	1995	
Aircraft, Hydrogen Fuel	2000	
(See also Table II-1)		
CONSERVATION BUILDING - EQUIPMENT:		
Electric Heat Pump, Single Family	1980	
Electric Heat Pump, Multi-family and Commercial	1980	
Gas Heat Pump, Single Family	1985	
Gas Heat Pump, Multi-family and Commercial	1985	
Ventilation Management	1980	2
Electric Heat Pump, Combined Air Conditioning	1980	
ELECTRIC AUTO:		
Battery-powered Auto	1985	
INDUSTRIAL CONSERVATION:		7
Specific Energy Reduction in Pulp and Paper Industry	1980	8
Specific Energy Reduction in Metals Industry	1980	8
Specific Energy Reduction in Iron and Steel Industry	1980	8
Ammonia Production from Coal	1985	
Solar to Process Heat	1985	
Atmospheric Fluidized Bed Industrial Boiler for Process Heat	1985	
Heat Management in Chemical Industry	1980	6
Heat Management in Other Industry	1980	6
Industrial Gas Heat Pump	1985	
(See Table II-1)		
RESIDENTIAL AND COMMERCIAL SOLAR HEATING:		
Solar to Space Heat, Single Family	1980	1
Solar to Space Heat, Multi-family and Commercial	1980	1
Solar Water Heat	1980	1
Solar Space and Water Heat	1980	1
Solar Air Conditioning	1985	

* Because of the large number of End-Use Technologies employed in the national data bases of the 15 participating countries, it is impractical to list all new technologies in a condensed summary. The reader is referred to the individual country reports for complete listings and detailed data.

** The START YEAR is defined as the year in which commercial units could be available for actual operation.

NOTES FOR TABLE II-2

1. Most solar heating systems modelled were required to have back-up systems of conventional types (oil, gas, electrical, heat pumps, etc.). The back-up systems could compete with each other to yield the most effective combination. The solar costs depend on country-specific insolation values.

2. Ventilation Management includes several specified packages of heat exchangers with recycle of recouped building exhaust heat.

3. Models of improved private autos projected gradual improvements with time in engine efficiency, the use of lighter weight structural materials and reduction in drag. Costs were usually expressed relative to conventional automobiles. Several countries did not model specific improvements but instead projected reductions in fuel demand per passenger mile in future years.

4. A variety of methanol-gasoline and methanol-diesel fuel mixtures were modelled, as well as pure methanol fuel. Cost differences in vehicles were minor.

5. A gradual shift in transportation modes from private auto to increased car pooling and mass transit is reflected in the national demand schedules for transport energy.

6. Gradual improvement in the heat management for specific industrial equipment is projected based on industrial advice as obsolete systems are replaced. Each industry based projections on particular devices, e.g. the recouping of waste heat for plant space heating and lower temperature process heating.

7. Most countries have included gradual improvements in the efficiency of industrial equipment in the demand schedules for industrial energy based on national analyses of conservation in industry.

8. A few countries have projected efficiency improvements for specific industries based on industry data.

ANNEX III

OVERVIEW OF SYSTEMS ANALYSIS RESULTS

The quantitative material presented in this Annex represents a summary of a large body of data generated in the energy systems analysis effort associated with the strategy project. The guiding principle in determining the material to be included has been whether the data provided new insights towards understanding the relative roles of new technologies in future energy systems.

The primary focus of the energy systems analysis was the performance of new energy technologies towards achieving goals of least cost energy suppply and maximum reduction of dependence on imported oil.

Framework for Evaluating Technologies

In order to assess the new technology potential, it was necessary to develop analytical tools and data that described some possible configurations of future energy systems in their entirety. The conceptualization of the analytical framework required that ranges of assumptions describing the future be made and that the analysis examine alternative energy policies that may be pursued by governments. These assumptions have been described in Chapter II.

There exist a rather large set of future energy policies that might be adopted to guide the development of new energy technologies. This study addressed three major policy positions that seem to have the greatest influence on new technologies. These policy areas are:

1. An energy policy preference for least cost energy supply or for security of energy supply;
2. An energy policy preference for maximum promotion of renewable resources;
3. An energy policy to accelerate the RD & D efforts for development of a wide range of new technologies.

The energy systems analysis used sixteen different scenarios that were designed to examine these energy policies under the various assumptions that might exist. Each scenario was designed to capture information on the behaviour of 27 categories of generic technologies which eventually included more than 100 specific technology processes.

Summary of Total Energy Systems Results

The analytical structure was formulated such that two key indicators could be used to focus the energy systems results. The first indicator was the discounted energy system cost. The second indicator was cumulative oil imports. These two indicators provide a relative measure of the trade-offs of energy policy alternatives to meet the perceived future needs.

In Figure III-1, the import values and the discounted system costs for a selected set of scenarios are presented. The values presented are for the aggregation of the fifteen IEA countries. Because this represents the first time that such aggregated data has been tabulated, the magnitudes of the numbers are not easily comparable to other known information. Therefore, the real focus should be placed on the relative values and the differences that exist between cases.

The year-by-year composition of the primary energy inputs has been presented for four scenarios in Chapter II. Despite the many internal variations of new technologies and the differing means for supplying end-use requirements in the different scenarios, the aggregation of primary energy supplies for fifteen countries presents a picture that does not drastically change among scenarios. In Figures III-2 and III-3, the primary energy requirements over the 1980-2020 time period are shown for the PS-1 and the SP-4/PREM-2 cases. The major change that can be detected from these graphs is the reduction of liquids in the SP- 4/ PREM- 2 case. The substitution of small amounts of energy inputs from each of the other five classes of energy source to displace the liquids does not create major changes in the general picture.

A more significant change in the primary energy input picture over time is illustrated in the maximum renewable energy scenario (RP-4) shown in Figure III-4. The significant increase in renewable energy does create a shift in the energy picture particularly during the 1980-2020 period.

Illustration of New Energy Technology Penetration in Different Scenarios

In each of the 16 scenarios for each country, the main attention is given to the performance of new energy technologies across the different cost/security trade-off situations, the accelerated technology situations, and the sensitivity situations. Through the study and evaluation of these comparisons, through the understanding of the underlying factors that contributed to the level of market penetration of individual technologies, and through the comparison of scenario needs versus technology behaviour, one can develop the priorities and the strategy for future RD & D activities.

An illustration of how technology performance can be evaluated is shown in Figure III- 5. This figure, based on the results of one country only, shows the cost/security trade-off curve for the non-accelerated technology cases. The minimum cost case (PS- 1) is used as a reference point for the new technology performance. At the bottom of the chart, a listing of the new energy technologies that entered the case solution is given. The list is divided into time periods so that one can see when a particular technology was competitively selected in the minimum cost case.

On the trade-off curve, the cases (SP- 1/PREM- 1, SP- 1/PREM- 2, and SP- 1/PREM- 3) are shown based on their cumulative oil import and discounted energy system cost indicators. At the top of the chart, a listing is shown for the changes in technology performance occuring in each case relative to the minimum cost case. There are two types of changes: (1) a technology that was not used in the minimum cost case may be now included in one of the trade-off cases; or (2) a technology that entered the minimum cost case at a later date may have an earlier time of introduction in a trade-off case.

As an example, enhanced gas recovery enters the minimum cost case in 1995, but in the SP- 1/ PREM- 1 case, it begins to be used in 1985. This observation gives an indication of the attractiveness of a new technology if a country was willing to pay a higher cost for the displacement of imported oil. Also, it can be seen that the wind central electric technology enters the SP- 1/PREM- 1 case in 1995 whereas it did not enter the minimum cost case.

New Technology Performance

In this section, the performance of each generic technology category is discussed in terms of certain measurement indicators. The indicators are the time of penetration relative to time of availability for use, the number of countries using the technology, the magnitude of the energy contribution over time in selected cases, and significant variation of performance in sensitivity cases.

A relative overview of the energy impact of the different technologies and the general applicability to countries is depicted in Figure III- 6. This Figure shows the relative energy contribution in the year 2010 for the two main scenarios, PS- 1 and SP- 4/PREM- 2, used in the strategy development.

Summary tables of the estimated energy contributions by ten year intervals for each category of new technology are interspersed through the section in Tables III- 7 to III- 11. The values are given for the PS- 1 case and the SP- 4/PREM- 2 case.

In the discussion of the energy impacts, the minimum cost case (Oil Price Schedule B) is given first. Comparisons to these values are made based on the important changes occurring in the other cases.

END-USE TECHNOLOGIES

Automotive Transportation Systems
See Figure III- 11

Although a variety of new technologies for the transportation sector were examined, it turned out that only a few of these technologies had any significant impact on the energy system. These technologies were mostly new automobile engines such as the otto cycle, the diesel, or the sterling engine. These technologies were included in the analysis for only a few countries, while the remaining countries, lacking sufficient data on the technology, attempted to incorporate their impact within the energy demand schedules.

The contributions of this technology are dominated by two countries. The energy contributions were 30 Mtoe/year in 1990, 170 Mtoe/year in 2000, and 370 Mtoe/year in 2010. These contributions include both the fuel inputs to the device as well as the conservation savings of the device relative to a conventional technology.

For the accelerated security case, the contributions were roughly doubled during the first 20 years of the analysis, but fall to lesser values in the post-2000 period due to the increased availability of more economically attractive alternative fuels. The energy impacts are relatively unaffected by the higher oil price cases.

Alternative Transportation Fuels
See Figure III- 11

This category includes liquid fuels substituting for conventional petroleum products such as methanol from natural gas, gasoline from methanol, and gasoline from natural gas. This technology had a designated availability in the 1980-1985 time frame for a few countries but was not generally available to most countries until around the year 2000.

These processes are implemented in 10 countries in the minimum cost case and in 14 countries in the accelerated cost/security case. The energy contribution values are dominated by two countries before the turn of the century and dominated by one country after the turn of the century. The contributions are 20 Mtoe/year in 1990 and 2000, 60 Mtoe/year in 2010, and 200 Mtoe/year in 2020 in the minimum-cost case.

In the accelerated security case, the earlier availability and more rapid penetration schedule lead to substantial increases to 85 Mtoe/year in 2000 growing to 380 Mtoe/year by 2020. These technologies do not show additional contributions in the higher oil price cases primarily due to the fact that they are already being implemented near their upper bounds in the associated reference case. The contributions are severely constrained to about 20 per cent of their reference values in the limited fossil cases.

Conservation Building (Equipment)
See Figure III- 11

Numerous end-use devices were represented by individual countries that would improve the efficiency of energy use in residential and commercial buildings. Typical examples were the use of electric heat pumps, gas heat pumps, and low temperature heat storage. Although a large number of different devices were characterized, it was not feasible for all of the end-use devices that are undergoing research at this time to be included in the analysis. Therefore, these contributions are representative of the impact that may be achieved by more aggressive use of the new equipment in buildings.

This technology category achieved penetration in fourteen countries in the minimum-cost case and in fifteen countries in the accelerated-security case. Depending upon the type of device, the economic competitiveness began as early as 1980 in some countries and introduction of other devices was continued through the year 2010 for other countries. While the contributions were substantial in most of the countries, it should be noted that the aggregate totals were dominated by one country.

The energy contribution is represented as the fuel input to the using device plus the energy saving of the device relative to the comparable conventional device. The contribution of both existing and new technologies was 90 Mtoe/year in 1990,

210 Mtoe/year in 2000, and 350 Mtoe/year in 2010. The energy contribution increased between 10-20 per cent per year in the various time periods in both the accelerated-security case and the maximum renewables case (RP-4). An additional 10-20 per cent increase in contribution was observed for each time period in the limited fossil cases.

Electric Auto
See Figure III- 11

The use of electricity to power automotive vehicles for road transport was investigated in the analysis. In general, the market application was restricted to the urban transportation needs which represents the most favourable electric auto market.

The technology was designated as available in 1985 and was found to be competitive by 1990 in four countries. The electric auto eventually became competitive in six countries in the minimum-cost case and in ten countries in the accelerated-cost/security case. Despite the number of implementing countries the energy contribution was small and was dominated by one country. The values were 2 Mtoe/year in 2000, 6 Mtoe/year in 2010, and 10 Mtoe/year in 2020 in the minimum-cost case.

In the accelerated security case, the electric auto showed improved penetration by 2000 but did not exceed 20 Mtoe/year at any time in the analysis. The most significant situation for the application of the electric auto occured in the limited fossil cases. In these situations, the electric auto technology was chosen by an additional country that had not used it in any other case and the implementation level was quite significant. The energy contributions in the limited fossil cases were of the order of 50 Mtoe/year in 2010 and from 120 to 200 Mtoe/year in 2020.

Industrial Conservation
See Figure III- 11

In the analysis, this category was separated into two components; industrial heat management applications that could be common to a wide variety of industries; and industrial process specific applications that were applicable only to a particular industry. These technology devices included such items as medium temperature heat from binary cycles, industrial cogeneration, medium temperature heat from solar energy and several individual processes for energy intensive industries. These devices were not explicitly included in all countries nor was it practical to represent all of the possible energy saving techniques that might be implemented within industrial applications. However, these values should provide a measure of the potential energy impact of these technologies within the framework of this analysis.

These technologies were implemented in all countries that explicitly included them in their systems representation. The energy contributions were 115 Mtoe/year in 1990, 200 Mtoe/year in 2000, 300 Mtoe/year in 2010, and 400 Mtoe/year in 2020 in the minimum-cost case. These values increased to 185 Mtoe/year in 1990, 320 Mtoe/year in 2000, 410 Mtoe/year in 2010, and 485 Mtoe/year in 2020 in the accelerated security with premium cases. The increases in oil prices, the limited nuclear and limited fossil cases have very little effect on these values.

Residential and Commercial Solar
See Figure III- 9

A wide variety of solar systems were characterized for individual countries for use in space heating, space cooling, and water heating of both residences and commercial establishments.

These systems were augmented with conventional fuel systems, where necessary, to provide auxiliary or back-up energy. The availability date was specified as 1980 for the technology but these systems did not become competitive until the 1985-1990 time period. The technology was selected for implementation in ten countries in both the minimum-cost case and the accelerated-security case. However, the energy contribution of this technology was dominated by the contributions from two countries. The energy contribution in the minimum-cost case started with rather small implementation of 4 Mtoe/year in 1990 but progressed steadily throughout the time period. The contribution was 16 Mtoe/year by 2000, 35 Mtoe/year in 2010, and 50 Mtoe/year in 2020.

In the accelerated cost/security case, the contribution increased to 40 Mtoe/year in 2000, 75 Mtoe/year in 2010, and 105 Mtoe/year in 2020. However, the contributions were again dominated by two countries while the other countries showed very small increases. The contributions were relatively unchanged in the higher oil price cases except to yield slightly higher contributions in the 1990 time period. The limits of feasible applications of this technology were further illustrated by the fact that in the limited fossil and limited nuclear cases there were no significant changes in the contribution.

PRODUCTION

Enhanced Gas Recovery
See Figure III- 8

This technology represents the recovery of natural gas from tight geological formulations, which would involve the use of advanced drilling techniques, hydrofracturing, or explosives to free the trapped gas within the formations. Due to the different types of geological formations, two different cost categories for recovery were established with varying amounts of potential recovery at those cost levels.

The technology was given an availability date of 1985 but it was not until 1995 that the gas became competitive. The enhanced gas technology was employed in only one country in the analysis. However, the energy contribution for that one country was judged to be very large with a value of 70 Mtoe/year in 2000, 110 Mtoe/year in 2010, and 150 Mtoe/year in 2020.

The accelerated security case led to earlier production of the gas but the general pattern of contribution was relatively unchanged. These contributions were relatively unaffected by higher oil prices. A significant increase was observed in the limited nuclear cases, while a significant reduction occurred in limited fossil cases.

Enhanced Oil Recovery
See Figure III- 8

The additional recovery of conventional domestic oil through such means as thermal processes and chemical processes was included in this category. Because of the differences in reservoir characteristics, three separate extraction costs were established for varying types of oil. An upper bound estimate on the total recoverable resource at each cost level was specified.

The availability of enhanced oil recovery was specified as 1980 and was judged economic in two countries in 1980. There were only two countries to select this technology primarily due to the non-availability of oil resources in most of the other countries. The lower cost resource was selected in the first time period and increasing quantities were further selected as the projected world oil price increased. The pattern established in the analysis was to use the product as rapidly as possible. Consequently, in one country, the total recoverable resource was judged to have been completely used up shortly beyond the year 2000. In the second country, the recovery hit a peak around the year 2000 and declined thereafter.

In the minimum-cost case, the energy contribution was 140 Mtoe/year in 1990, 120 Mtoe/year in 2000, and declining to 45 Mtoe/year in 2020. In the accelerated-security case, the contribution was 50 Mtoe/year in 1990, 175 Mtoe/year in 2000, and declining to 30 Mtoe/year in 2020. There were relatively minor changes in these values in the sensitivity cases involving higher oil prices, limited fossil application, or limited nuclear application.

Geopressurized Methane
See Figure III- 8

Geopressured hydrothermal resources are hot water reservoirs containing methane which are trapped under high pressure. Based on the analysis, this potentially large resource was thought to be economically recoverable only in one country.

Initial quantities of this resource are judged to be available by 1990 but were not competitive until the year 2000. In the minimum-cost case, the energy contribution was 10 Mtoe/year in 2000 and reaches a peak contribution of about 40 Mtoe/year in 2010. The decline in contribution in the later years of the analysis was due to a shift in end-use requirements for gas rather than a resource limitation.

In the accelerated-security case, the contributions are the same through 2010, but there was an increase to 70 Mtoe/year by 2020. Even larger quantities, about 30 per cent greater, of the resource were desired in the 2020 period in the limited nuclear case. In the limited fossil case, there is no contribution from this resource.

Geothermal/Hydrothermal
See Figure III- 9

A geothermal power plant of 110 MWe size using wet steam from hydrothermal reservoirs was used in this technology category. The technology was limited to those countries and to those areas within the countries where an appropriate resource exists.

The date of availability was specified as 1980 for four countries and was found to be economically competitive in all four countries at that early date. The resource was judged to be eventually available in six countries and successfully competed in the energy system in all six countries.

Because of resource limitations, the energy contribution in most of the countries was rather small, resulting in the aggregate continuation being dominated by two countries. In the minimum-cost case, the energy contribution was 6 Mtoe/year in 1990, 13 Mtoe/year in 2000, 40 Mtoe/year in 2010, and 65 Mtoe/year in 2020. These contributions were doubled up to the year 2000 in the accelerated-security case while being marginally higher in the post-2000 years. These latter values were only about 5 per cent below the specified geothermal potential achieved in the renewables case (RP- 4). Similar values were observed in both the limited nuclear and limited fossil cases.

Geothermal Hot Dry Rock
See Figure III- 9

An advanced geothermal technology using the energy from the deep rocks within the earth's surface was investigated for both power plant applications and district heating applications. This resource was presumed to be available in 1990 in limited quantities at low cost but available in large quantities only at much higher costs.

The technology was found to be competitive in six countries although the quantities produced from hot dry rock were extremely limited in all time periods. The energy contribution rose only to 5 Mtoe/year by 2020 in all the scenarios except the renewables case, where it achieved a value of 8 Mtoe/year in 2020. In the renewables case, the technology was implemented at very small levels by an additional two countries.

Ocean Power
See Figure III- 9

Three technological processes were represented by different countries in the ocean power category. These processes were: a wave central electric power plant of small size; an ocean thermal gradient power plant of medium size; and a tidal electric plant of large size.

The availability date for the technology was designated as 1990. Limited applications of the technology were found to be economically competitive in one country in the year 2000 and in a second country in 2010 in the minimum-cost case. The total contribution in this case did not exceed 10 Mtoe/year.

The ocean power technologies did demonstrate better potential in the limited fossil and limited nuclear cases where the technology achieved a value of 20 Mtoe/year in 2010 and 30 Mtoe/year by 2020. Because of its generally uneconomic nature, until the very late time periods, the technology did not show improved capabilities in the accelerated technology cases.

Shale Oil and Tar Sands
See Figure III- 8

The extraction of oil from shale by surface retorting and through in-situ processes as well as the extraction and processing of tar sands through a variety of methods were included in this category.

The technology availability date was 1980 for tar sands and 1985 for shale oil. Tar sands were found to be economic in one country in its initial time period. Shale oil was not judged to be economic until 1995 in the minimum cost case. Only one country selected tar sands for implementation and only one country had significant implementation of shale oil. Two additional countries did have extremely minor quantities of shale oil in the latter time periods of the analysis.

In the minimum-cost case, the total contribution in this category was 40 Mtoe/year in 1990, 120 Mtoe/year in 2000, 190 Mtoe/year in 2010, and 230 Mtoe/year in 2020. In the accelerated-security case, increases occurred in the 1990 period to 70 Mtoe/year and 140 Mtoe/year by 2000. The post-2000 contributions were comparable to the minimum cost scenario. The higher oil price cases had no affect on the energy contribution as these technologies were being implemented to their upper bounds in the reference cases. There were very small changes in energy impact in the sensitivity cases for limited fossil and limited nuclear.

Solar Electric
See Figure III- 9

In the analysis process, separate categories for dispersed solar electric and centralized solar electric were analysed. These processes included the decentralized solar photovoltaic process, a central solar thermal electric technology, and a central solar photovoltaic power plant.

Although the availability date of the technology was specified by 1985, only the decentralized version was found to be economically competitive in only one country by the year 2000. An additional three countries found the technology to be competitive in the post-2010 time frame. The energy contribution was extremely small until the year 2020 when a value of 5 Mtoe/year was achieved in the minimum-cost case.

In the sensitivity cases, the number of implementing countries increased to seven but the energy contribution only increased to 15 Mtoe/year in the limited nuclear and limited fossil cases. A measure of the upper potential of these technologies was illustrated in the accelerated renewables case where the maximum contribution was 30 Mtoe/year by 2020. The deployment of this technology was limited in many countries by poor solar insolation.

Wind Power
See Figure III- 9

Two sizes of wind energy devices were examined in the analysis: a large central station wind turbine of 2 MWe size and a local small wind electric generator of 20 KWe.

The availability date was specified as 1985 and one country found the small wind generators to be economic in that time period. Only four countries implemented the technology in their energy futures in the minimum-cost case with one of these countries not finding the technology economic until 2020. The energy contribution is rather limited with an aggregate contribution of 5 Mtoe/year by 2000 and growing to only 8 Mtoe/year by 2020.

In the accelerated security case, wind power performed to a lesser extent as it was replaced by the earlier availability of more economic technologies. A measure of the maximum potential of wind power that might be implemented under favourable circumstances was obtained in the maximized renewables case. In this situation, wind power contributed 50 Mtoe/year in 2000 and 85 Mtoe/year by 2020. Wind power achieved about one half of this potential in the limited fossil and limited nuclear cases. In all scenarios, the energy contribution of the technology was dominated by the impact in one country.

CONVERSION

Advanced Converter Reactors
See Figure III- 10

Fourteen different types of advanced converter reactors were characterized including power plants designed to produce only electricity, plants for gasification of coal, and plants for production of process heat. Some of these reactors will be available in some countries in the next few years, while other applications were judged to be not available until 1995-2000.

These reactors were found to be economic at varying time periods depending upon the individual country stage of development. The earliest implementation began in 1980 and additional implementation occurred in the mid-1990's for mostof the other countries. The advanced converter reactors was implemented in eight countries in both the minimum-cost case and the accelerated-security case. The energy contribution rose from 20 Mtoe/year in 2000 to 50 Mtoe/year in 2010 and 100 Mtoe/year in 2020 in the minimum-cost case. These values increased to 35 Mtoe/year in 2000, 60 Mtoe/year in 2010, and 120 Mtoe/year in 2020 in the accelerated security case. These higher values were also observed in the higher oil price situations. Even greater values were found in the limited fossil cases where the advanced converters contributed 95 Mtoe/year in 2010 and 170 Mtoe/year in 2020.

Breeder Reactor
See Figure III- 10

The liquid metal fast breeder reactor (LMFBR) was the only breeder reactor process characterized in this analysis. The characterization of the reactor included the development of the fuel cycle for reprocessing spent fuel and fabricating mixed oxide fuels. The cost of reprocessing existing reactor spent fuel is charged to the existing reactors rather than the LMFBR. The growth pattern for the reactors was dependent upon the availability of plutonium from existing reactors, and on fuel doubling time of the breeders.

The date of availability was specified as 1995 for several countries. The LMFBR was found competitive in 1995 in one country, competitive in an additional four countries by the year 2000, and eventually implemented in nine countries. After having become competitive, the LMFBR exhibits strong growth in the post-2000 time period in all implementing countries.

In the minimum-cost case, the energy contribution is 50 Mtoe/year in 2010 and 130 Mtoe/year in 2020. In the accelerated-security case, the implementation date is 1995 for several countries and the contribution grows from 15 Mtoe/year in 2000 to 80 Mtoe/year in 2010 and to 170 Mtoe/year in 2020. As the breeder reactor achieves maturity, it begins to replace existing converter reactors in the mix of electric generating capacities.

The higher oil price cases have little effect on the breeder implementation because it is already approaching maximum penetration potential at the reference oil price. The breeder achieves about a 10 per cent increase in each time period in the limited fossil case while the contribution is cut in half in the limited nuclear case.

Coal liquefaction

See Figure III- 7

Coal liquefaction, the process of transforming coal into a liquid fuel, is represented by several different conversion processes for both hard and brown coal. These processes include hydrogenation, Fischer-Tropsch, Solvent Refined Coal, and methanol production by the Lurgi process.

The earliest date of availability was specified as 1985. The earliest date of competitiveness in the minimum-cost case was the year 2000. However, between 1985 and 2000 certain quantities of coal liquefaction were used in particular countries due to specified government commitments to build facilities. The technology was implemented at varying time intervals between 2000 and 2020 in eight countries. After the technology became competitive, the demand for its liquid product was very extensive with the energy contribution increasing from 40 Mtoe/year in 2000 to 470 Mtoe/year by 2020.

In the accelerated-security case, the competitive introduction of coal liquefaction began as early as 1990. Fifteen countries selected the technology for use. Again the growth was very large as the energy contribution increased from 110 Mtoe/year in 2000 to 650 Mtoe/year in 2020.

The only cases showing variation from this growth pattern were the limited fossil and limited nuclear cases. In the limited fossil case, as expected, the potential of coal liquefaction was muted by the requirements to reduce fossil fuel uses and the contributions are relatively minor. In the limited nuclear case, the initial introduction of coal liquefaction was slowed due to the fact that the coal was used for electric generating systems to replace the nuclear power and thus diverted away from liquefaction processes.

Coal Combined Cycle, Fluidized Bed Combustion

See Figure III- 7

The production of electricity from power plants of 1200 MW capacity, using a coal gasification process with fluidised bed combustion and a combined cycle operation of both gas and steam turbines, was incorporated in the analysis.

The availability date was specified as 1990. The technology became competitive in 1990 in one country but it was the year 2000 and beyond before achieving competitiveness in the remaining countries. The technology was selected by seven countries in the minimum-cost case and by eight countries in the accelerated-security case.

The energy contribution of the combined cycle/ fluidised bed combustion technologies in the accelerated-security case was 30 Mtoe/year in 2000, 50 Mtoe/year in 2010, and 45 Mtoe/year in 2020. The slow rate of growth in the latter time periods can generally be explained by the increase in nuclear power, which captures the bulk of the electricity market, and the preferred use of coal in other applications.

This technology had its greatest impact in the accelerated-security limited nuclear scenarios where its contribution increased to about 110 Mtoe/year in 2010 and 190 Mtoe/year in 2020. The technology experienced only minor increases in the accelerated technology situations and in higher oil price cases.

Fuel Cell

See Figure III- 7

The fuel cell technology examined in this analysis was a gas fuel cell of 26 MW size based on coal derived fuel. It is a technology designed for peaking power applications.

The technology was given an availability date of 1985 and was first competitive in 1990. The technology was selected by three countries in the minimum-cost case and by five countries in the accelerated-security case. Its energy contribution was extremely small in all time periods and in all cases. The maximum contribution from this technology in any situation did not exceed 1 Mtoe/year.

Fuels from Biomass
See Figure III- 9

The biomass energy category contained an extremely wide variety of processes that convert raw materials into a variety of energy forms. It was not possible to represent the entire spectrum of these energy forms in this analysis; however, ten different biomass processes were characterized and used by a variety of different countries.

Of the biomass processes included, varying types were found to be attractive for at least seven countries in both the minimum-cost case and the accelerated-security case. The failure of biomass to be implemented in the other countries was due primarily to limitations in land area available for biomass production.

In the minimum-cost case, the contribution of biomass was about 5 Mtoe/year in 2000, gradually growing to about 10 Mtoe/year in 2020. The accelerated-security case had the effect of approximately doubling these contributions in each of the time periods.

The technology did not show any significant increases in response to the higher oil price cases. The most significant application of biomass occurred in the limited fossil cases. The contribution went as high as 75 Mtoe/year in 2020 in the minimum-cost limited fossil case and as high as 125 Mtoe/year by 2020 in the accelerated-security limited fossil case. It would appear that biomass technology would be a significant alternative in a world that faced constraints on the application of fossil fuels.

Fusion Reactors
See Figure III- 10

Magnetic and inertial confinement fusion reactors were characterized for the analysis. However, all countries judged the technology not to be available until beyond the year 2020. Consequently, no energy contribution can be shown.

High Caloric Gasification
See Figure III- 7

The production of pipeline quality gas through transformation of coal by such processes as Slagging Lurgi gasifiers, hydrogasification, and nuclear steam gasification was examined in various countries. In countries where the existing pipeline system is limited, an additional cost for the construction of pipelines was added to the gasification system. This feature contributed to the lack of success of this technology in the analysis.

The technology was given an availability date of 1985 and was judged to be economic in one country by that time. The technology did not become economic in other countries until 1995 to 2005. The technology was selected for use in only four countries in both the minimum-cost case and the accelerated-security case. The energy contribution in the year 2000 was 25 Mtoe/year and grew to 80 Mtoe/year by 2020 in the minimum-cost case.

The technology did not show significant increases in any of the other cases. In several countries, this was due to the fact that other sources of pipeline-quality gas were available to meet the demands and the higher cost coal gasification technology was not able to successfully compete.

Low/Medium Caloric Gasification
See Figure III- 7

The transformation of both brown coal and hard coal into a lower-quality gas for use in industrial applications was investigated. This quality of gas can only be used in local or regional areas of industrial facilities which would be near the source of coal because transporting the gas over long distances is non-economic.

The date of availability was specified as 1980 but it did not become economically competitive in any country until the year 2000. The technology was selected for implementation in only two countries in any of the scenarios. The energy contribution was rather small even in these two countries with an impact of 10 Mtoe/year by 2010 and 20 Mtoe/year by 2020. These energy contributions were approximately the same in all scenarios except for the limited fossil situations where the impact was limited to small applications in only one country.

Magnetohydrodynamics (MHD)
See Figure III- 7

The MHD technology included in this analysis is a large-scale electric power plant where coal is burned at very high temperatures to produce an ionized gas which passes through a magnetic field to generate electricity. The technology is still in the early stages of development today.

The date of availability was specified as the year 2000 in a few countries and several years later for the other countries. The technology was judged to be economic in one country by the year 2000, but the scale of implementation was very small. The technology was not selected in the minimum-cost case until the year 2020 by any other country. The quantity of energy from this technology was less than 1 Mtoe/year by the year 2010 and less than 10 Mtoe/year in 2020.

In the accelerated-cost security case, two additional countries found the technology attractive in the year 2010. Even in these circumstances, the contribution was only 3 Mtoe/year in 2010 and 15 Mtoe/year in 2020.

As with the other coal-based electric producing technologies, MHD did show better performance in the limited nuclear scenarios where the contribution increased to about 8 Mtoe/year in 2010 and 35 Mtoe/year by 2020. The technology showed no increases due to the higher oil prices, which is a reflection of its uneconomic nature in many of the countries.

Nonfossil Hydrogen Production

See Figure III- 9

Two processes, hydrogen production by electrolysis and hydrogen production by thermal chemical splitting, were investigated in this analysis. Although these technologies were estimated to be available by as early as 1985 for the electrolysis and the year 2000 for the thermal chemical process, the technologies were not selected in significant quantities by any country.

The peak contribution was less than 1 Mtoe/year in 2020 in the minimum-cost case. The upper potential for the technology was illustrated in the accelerated-security limited fossil case where a contribution of 20 Mtoe/year in 2010 and 50 Mtoe/year in 2020 was achieved due to a large implementation in one country. It was observed that the technology was implemented by as many as nine countries in the sensitivity cases, but with the exception of the one country, the total energy contribution was extremely small.

Underground Coal Gasification

See Figure III- 7

The gasification of coal by processes of burning the coal in the ground and capturing the gaseous product through outlet wells was examined both for direct use of gas in industrial applications and for use with a combined cycle electric power plant.

The technology was judged available by 1985 in limited applications and by 1990 in expanded applications. The technology was found to be competitive in three countries by 1990 in very limited energy quantities. The technology was selected for implementation in a total of four countries in both the minimum-cost case and the accelerated-security case. However, the total energy contribution did not exceed 1 Mtoe/year in any time period.

In-situ coal gasification was found not to be economic even in the higher oil price situations. Even in the case with the additional premium being paid for the technology implementation, only very minor increases were observed to occur. There was no improved contribution from the technology in the limited nuclear cases.

Figure III-1. IEA ENERGY RD AND D STRATEGY PROJECT TOTAL CUMULATIVE OIL IMPORTS AND ENERGY SYSTEM COSTS (1980-2020) FOR 15 IEA COUNTRIES UNDER SEVEN SCENARIOS

	NON-ACCELERATED SCENARIOS	
	OIL IMPORTS btoe	TOTAL DISCOUNTED ENERGY SYSTEM COST $\$ \times 10^{12}$
PS-1	48. 0	14. 06
SP-1/PREM-1	39. 8	14.12
SP-1/PREM-2	38. 3	14.17
SP-1/PREM-3	37. 7	14. 20
SP-1	33. 5	15. 78
	ACCELERATED SCENARIOS	
PS-4	43. 2	13. 8
SP-4/ PREM-2	33. 1	14. 0

See Figure 6 for explanation of scenarios.

Figure III-2

IEA ENERGY RD & D STRATEGY PROJECT

PRIMARY ENERGY (FOSSIL FUEL EQUIVALENT)

Non-Accelerated Minimum Energy System Cost Scenario (PS-1)

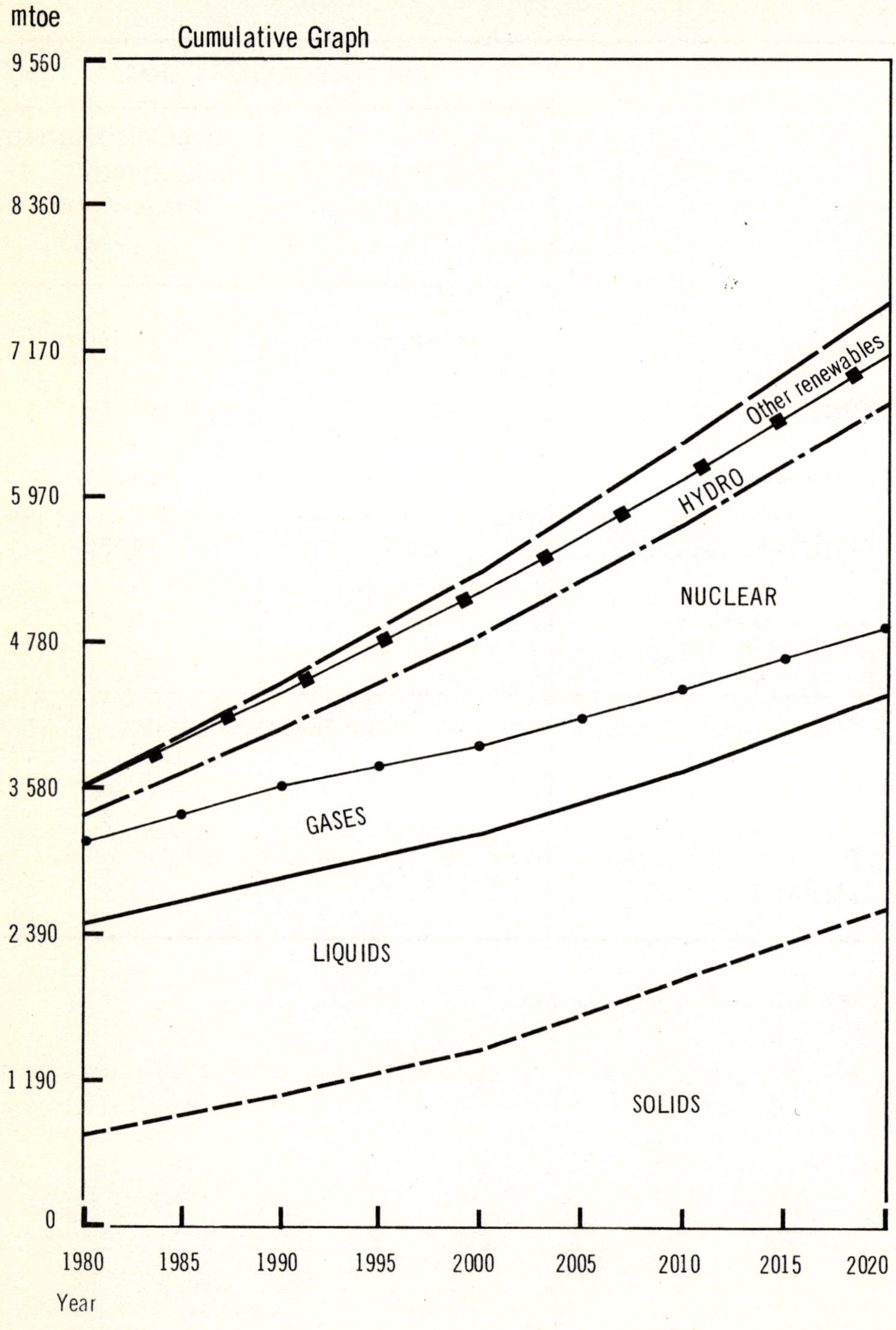

Figure III-3

IEA ENERGY RD & D STRATEGY PROJECT

PRIMARY ENERGY (FOSSIL FUEL EQUIVALENT)

Accelerated Security Scenario (SP-4/PREM-2)

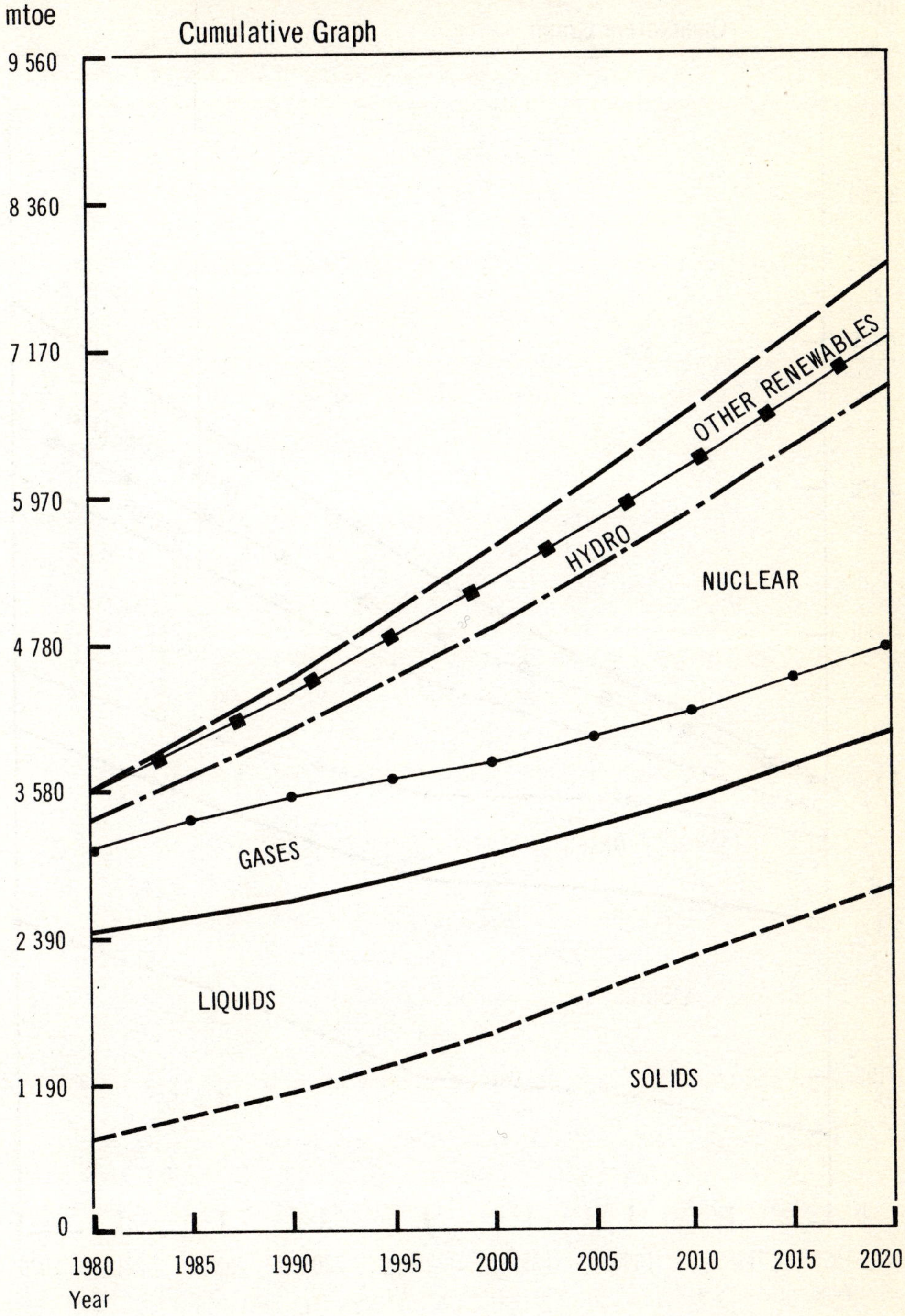

Figure III-4

IEA ENERGY RD & D STRATEGY PROJECT

PRIMARY ENERGY (FOSSIL FUEL EQUIVALENT)

Accelerated Maximum Renewables Scenario (RP-4)

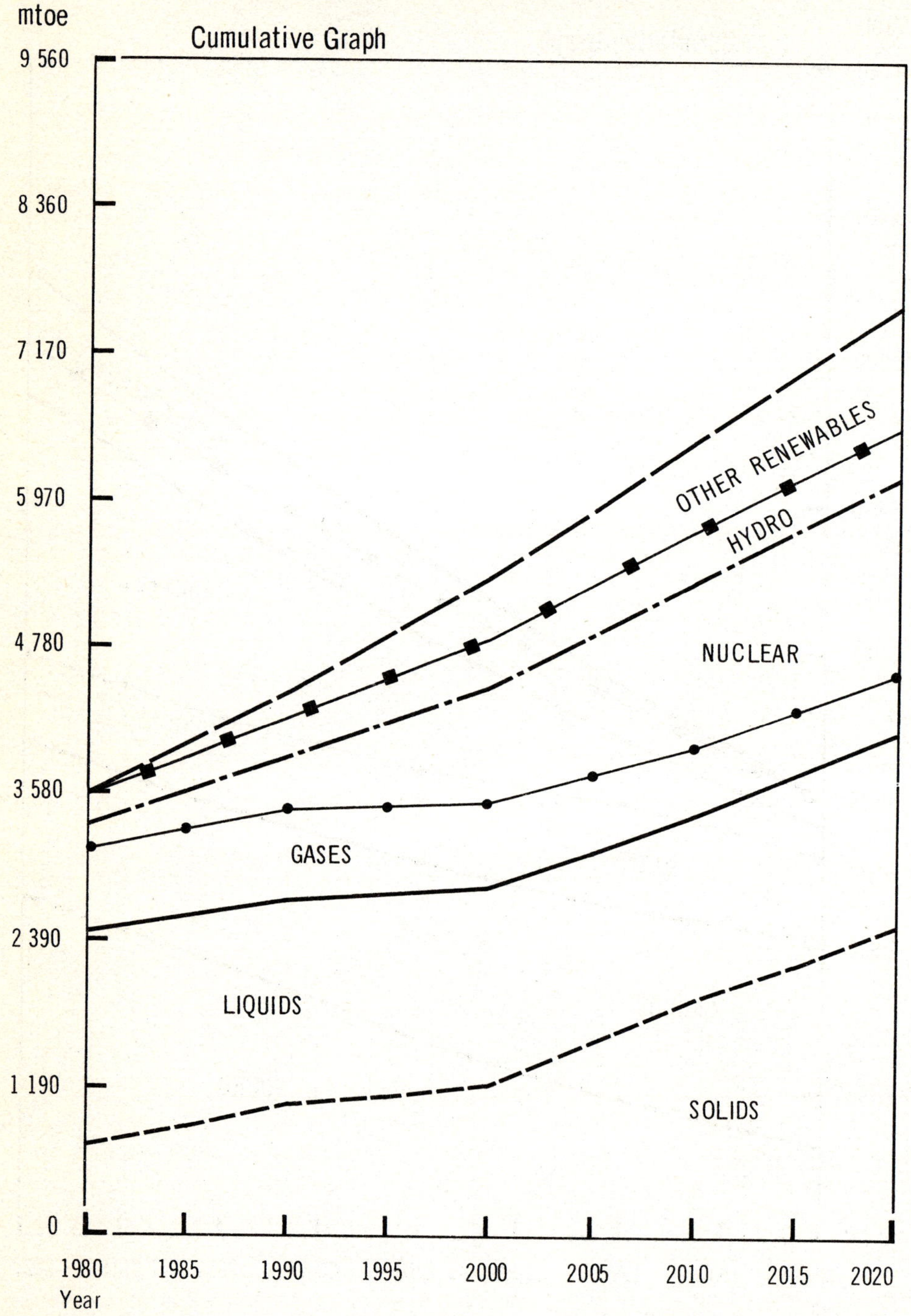

Figure III-5

IEA ENERGY RD & D STRATEGY PROJECT

TECHNOLOGIES USED IN ONE COUNTRY UNDER VARIOUS SCENARIOS

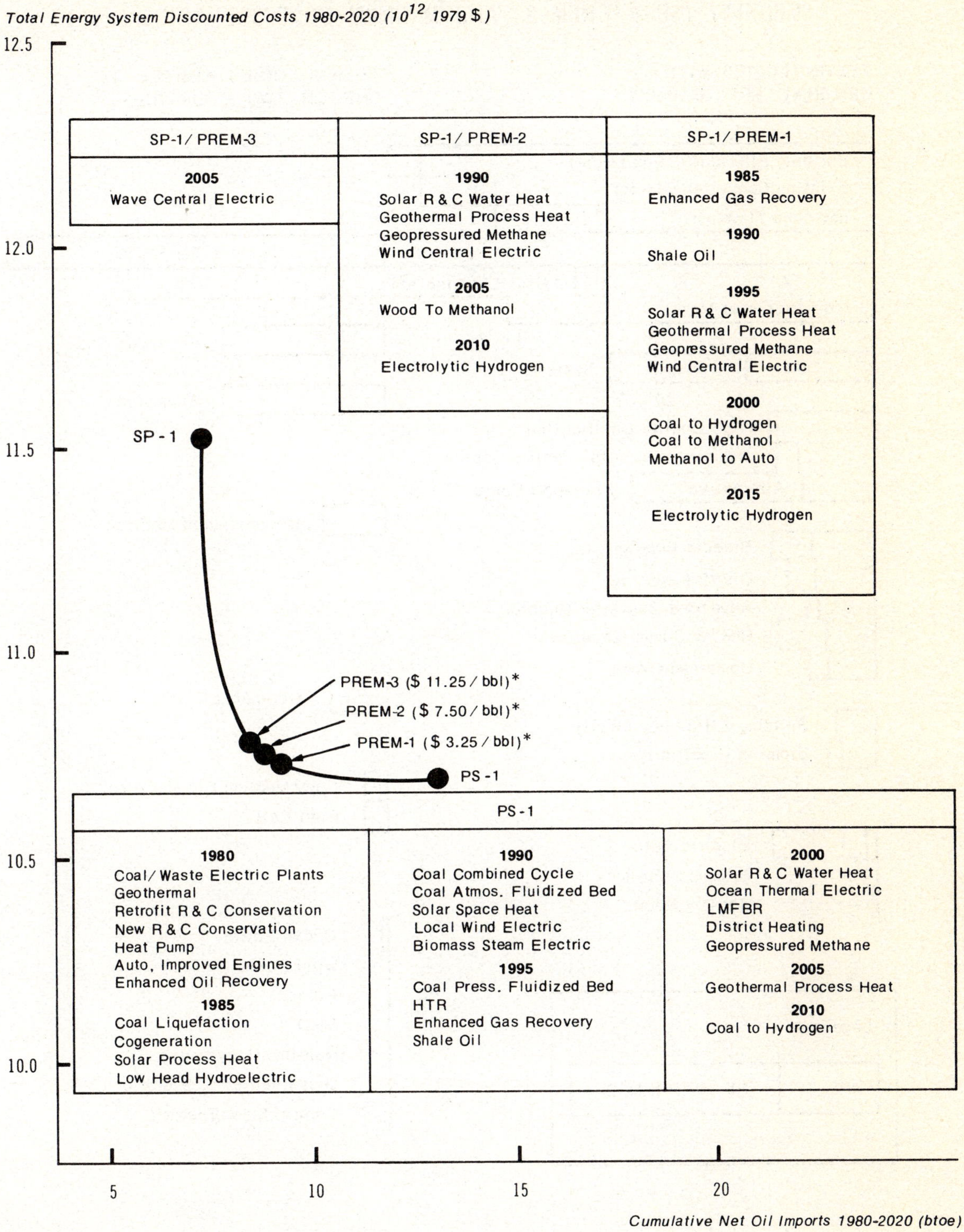

* See Ch.II.B.3 for explanation of these scenarios.

Figure III-6

IEA ENERGY RD & D STRATEGY PROJECT

ENERGY CONTRIBUTIONS OF TECHNOLOGIES IN 2010 IN THE ACCELERATED SECURITY (SP-4/PREM-2) AND MINIMUM COST SCENARIOS (PS-1)

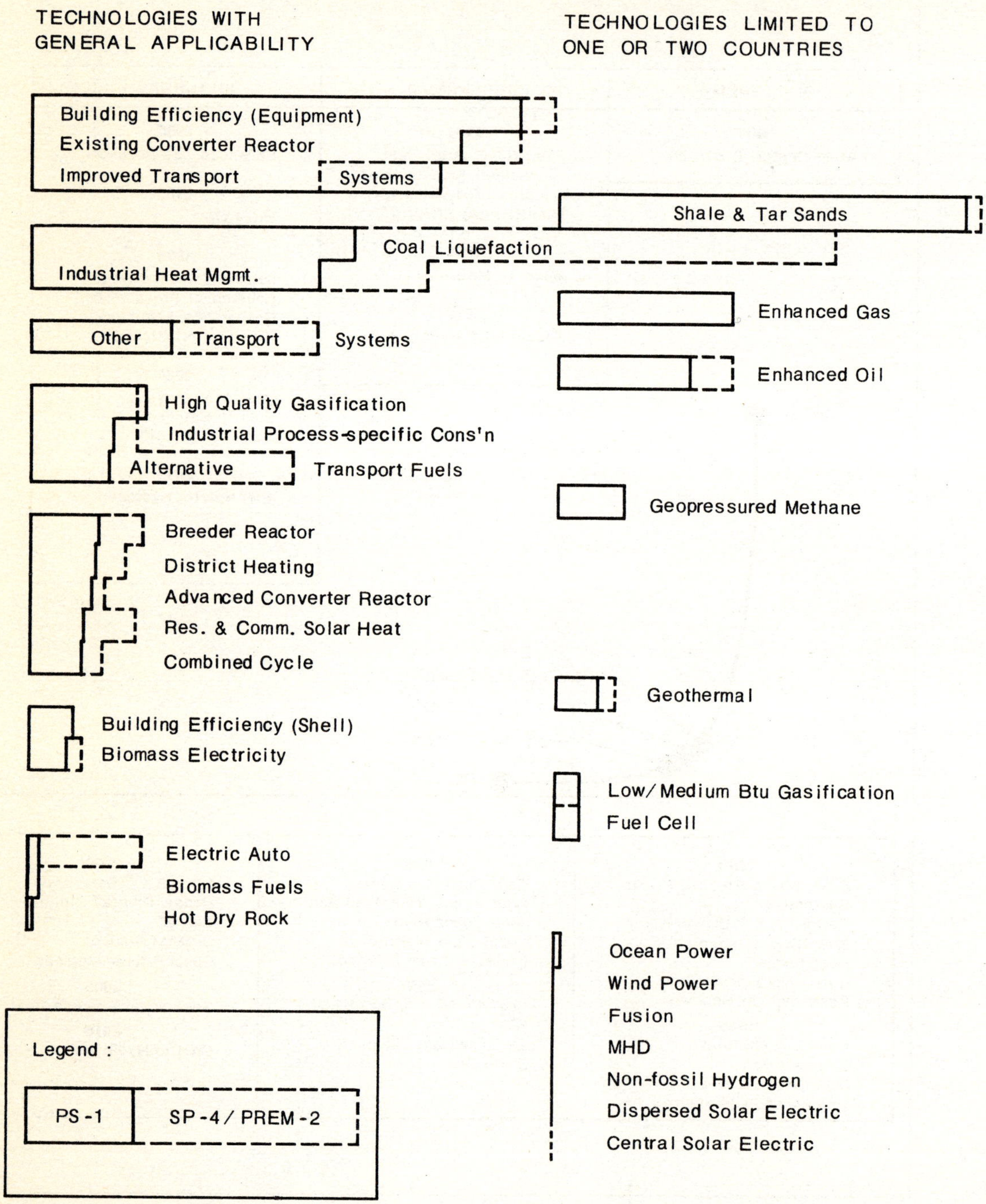

Figure III-7. IEA ENERGY RD AND D STRATEGY PROJECT CONTRIBUTION OF NEW COAL TECHNOLOGIES UNDER TWO SCENARIOS (Mtoe/year)

	MINIMUM SYSTEM COST (PS-1)				ACCELERATED SECURITY (SP-4/PREM-2)			
	1990	2000	2010	2020	1990	2000	2010	2020
Coal Liquefaction	5	39	223	467	11	109	420	646
High Calorie Gasification	13	24	53	78	13	31	48	73
Low/Medium Caloric Gasification	-	-	10	18	-	1	10	18
Combined Cycle	*	17	32	36	3	11	44	41
Fluidized Bed Combustion	2	14	14	3	2	16	8	4
Magnetohydrodynamics	-	*	*	8	-	*	3	16
Coal Gas Fuel Cell	*	*	2	*	-	*	*	*
In-Situ Coal Gasification	*	*	1	2	*	*	2	3

* Less than 1 Mtoe/year.

Figure III-8. IEA ENERGY RD AND D STRATEGY PROJECT
CONTRIBUTION OF NEW OIL AND GAS TECHNOLOGIES
UNDER TWO SCENARIOS (Mtoe/year)

	MINIMUM SYSTEM COST (PS-1)				ACCELERATED SECURITY (SP-4/PREM-2)			
	1990	2000	2010	2020	1990	2000	2010	2020
Enhanced Oil Recovery.......	144	116	75	45	50	175	107	26
Enhanced Gas Recovery......	-	73	110	146	58	73	110	129
Shale Oil/Tar Sands	36	115	190	231	86	169	240	273
Geopressurized Methane	-	9	43	26	-	9	43	70

Figure III-9. IEA ENERGY RD AND D STRATEGY PROJECT
CONTRIBUTION OF NEW RENEWABLE TECHNOLOGIES
UNDER TWO SCENARIOS (Mtoe/year)

	MINIMUM SYSTEM COST (PS-1)				ACCELERATED SECURITY (SP-4/PREM-2)			
	1990	2000	2010	2020	1990	2000	2010	2020
Residential and Commercial Solar	4	17	37	48	*	43	74	105
Wind Power	2	5	8	8	2	3	5	8
Ocean Power	-	2	9	9	-	-	6	8
Fuels From Biomass	2	5	7	10	4	7	9	19
Solar Electric	-	*	*	4	*	*	*	4
Non-Fossil Hydrogen Systems .	-	-	*	*	-	*	1	8
Geothermal Hydrothermal	7	13	39	64	15	28	48	66
Geothermal Hot Dry Rock	*	1	3	5	*	1	3	6

* Less than 1 Mtoe/year.

Figure III-10. IEA ENERGY RD AND D STRATEGY PROJECT
CONTRIBUTION OF NEW NUCLEAR TECHNOLOGIES
UNDER TWO SCENARIOS (Mtoe/year)

	MINIMUM SYSTEM COST (PS-1)				ACCELERATED SECURITY (SP-4/PREM-2)			
	1990	2000	2010	2020	1990	2000	2010	2020
Advanced Converter Reactors..	7	22	53	102	15	34	63	118
Breeder Reactors	-	8	50	128	-	18	84	172
Fusion[1]	-	-	-	-	-	-	-	-

1. Not Available Until Post-2020.

Figure III-11. IEA ENERGY RD AND D STRATEGY PROJECT CONTRIBUTION OF NEW CONSERVATION TECHNOLOGIES UNDER TWO SCENARIOS (Mtoe/year)

	MINIMUM SYSTEM COST (PS-1)				ACCELERATED SECURITY (SP-4/PREM-2)			
	1990	2000	2010	2020	1990	2000	2010	2020
Conservation-Building Equipment	94	210	352	400	112	234	374	418
Automotive Transportation Systems	31	170	370	262	73	314	264	113
Alternative Transportation Fuels	19	20	59	198	10	88	192	382
Electric Auto	-	2	6	10	1	7	16	22
Industrial Conservation	114	197	302	403	186	321	410	485